Related volumes

Brickwork 1 and Associated Studies
Brickwork 2 and Associated Studies } Harold Bailey and David Hancock

Other title of interest

The Skills of Plastering, Mel Baker

BRICKWORK 3
AND ASSOCIATED STUDIES

Harold Bailey

Sometime Senior Lecturer
Stockport College of Technology

David Hancock

Senior Lecturer
Stockport College of Technology

Second Edition

First published 1979 by
THE MACMILLAN PRESS LTD
Houndmills, Basingstoke, Hampshire RG21 2XS
and London
Companies and representatives
throughout the world

ISBN 0-333-51957-4

A catalogue record for this book is available
from the British Library.

Printed in Malaysia

First edition reprinted twice
Second edition 1990
10 9 8 7 6
00 99 98 97

CONTENTS

PREFACE

This series of three volumes is designed to provide an introduction to the brickwork craft and the construction industry for craft apprentices and all students involved in building. All too often, new entrants to the construction industry are expected to have a knowledge of calculations, geometry, science and technology irrespective of their previous education. It is the authors' aim to provide a course of study which is not only easily understood but is also able to show the relationship that exists between technology and associated studies.

The construction industry recognises that the modern craftsman, while maintaining a very high standard of skills, must be capable of accepting change – in methods, techniques and materials. Therefore it will be necessary for apprentices to develop new skills related to the constant advancements in technology.

This third volume concludes the complete Craft Certificate course for the City & Guilds of London Institute, and includes the many other areas of work in which the craftsman is required to demonstrate his ability.

To become a highly skilled technician in the modern construction industry, the apprentice should recognise that physical skills must be complemented by technology, and that planned methods of construction must be used in all work situations.

The apprentice and young craftsman will be able to appreciate the diversity of the bricklayer's craft, and to relate his own abilities and ambitions to the immense scope offered by today's construction industry.

H. BAILEY
D. W. HANCOCK

ACKNOWLEDGEMENTS

The authors wish to acknowledge the assistance and cooperation of: The Clay Pipe Development Association Ltd, for figures 5.19, 5.20 and 5.21; S.G.B. Scaffolding (Great Britain) Ltd, for figures 6.6 to 6.21, 6.29 and 6.33; Hilti Ltd for figures 6.26, 6.27 and 6.28; The Brick Development Association, for figure 6.30; Walter Somers (Materials Handling) Ltd, for figures 6.40, 6.41 and 6.42.

1
SAFETY ON SITE

In this chapter we review the two main aspects of safety in construction. First, we look at the safety of workers on, and visitors to any construction operation. Secondly, but of equal importance, is the safety of the public in the area of the site, in particular the protection of pedestrians.

SAFETY ON SITE

The Construction Industry has always had an exceptionally high accident rate. In the year 1987-1988, 157 operatives lost their lives, 3624 suffered major accidents (see below) and around 20,000 reportable accidents (those involving at least three days off work) occurred. How many accidents occur involving less than three days off is anyone's guess, certainly measured in hundreds of thousands. Figures published by the Health and Safety Executive show that over 1000 operatives have lost their lives in the last eight years.

Such was the concern in the Construction Industry that 1983 was designated 'Site Safety Year', but regrettably even then the abysmal record did not really improve nor has it since.

The word 'accident' is defined as 'something that cannot be avoided', but since a great many of the so-called accidents could certainly have been avoided, it is obvious that the safety record will not improve until the operative on site wakes up to the potential dangers, shakes off his macho image and starts to take a responsible attitude towards his own safety and that of his workmates.

It is not proposed to list typical 'accidents' here, it would take too much space. Suffice it to say that using inferior scaffolds or poor timbering to trenches, lifting heavy objects by bending from the waist, not wearing safety boots (resulting in nails in the feet, or crushed feet), being too tough to wear a hard hat etc. is asking for trouble. And the figures I've previously given surely prove this to be so.

ACCIDENT PROCEDURE

The Reporting of Injuries, Diseases and Dangerous Occurrences Regulations 1985 (RIDDOR) came into force on 1 April 1986. Full details and guidance on these and other aspects of the Regulations are to be found in Guidance booklet HS(R)23 which is available from HMSO. The following is a brief introduction to these Regulations:

> Whenever any of the following events occurs it must be reported in writing within 7 days on form F2508 to the enforcing authority, usually the Health and Safety Executive, and events a, b and c notified as quickly as possible, that is, by phone.
>
> (a) The death of a person resulting from an accident at work.
>
> (b) Any person suffering a major accident such as a large bone fracture, amputation, loss of sight etc. or any other injury which results in hospital admission for a period in excess of 24 hours.
>
> (c) A dangerous occurrence such as failure or overturning of a lift, crane, hoist, scaffold over 5 m high, or the collapse of a structure or part of a structure.
>
> (d) A person at work being incapacitated for more than three days as a result of an injury caused by an accident at work.
>
> (e) The death of an employee if this occurs within a period of one year of the reported injury and can be solely attributed to the accident.

AFTER AN ACCIDENT

Where an accident has occurred and an injury sustained, the procedure to be followed depends on the circumstances. The obvious starting point is to tend the injured person, but firstly make sure it is safe to do so. For example, if an operative is lying unconscious and bleeding at the foot of a scaffold – first look upwards and assess if the situation is dangerous and likely to cause further injury to the injured operative or to yourself. Is the operative unconscious in a closed room in a gas-filled atmosphere? Is the operative lying unconscious holding an electric drill which is still connected to the mains? The procedure then is:

(1) Attend to the injured person if it is safe to do so.
(2) Ensure no-one else is injured for the same reason.
(3) Send for the qualified first-aider.
(4) Determine the cause of the accident.
(5) Obtain statements from witnesses while the incident is fresh in their minds.
(6) Notify the Health and Safety Executive as previously explained.
(7) Record the incident in the accident book.

KEEPING RECORDS

A record must be made and kept of all reportable injuries and dangerous occurrences, each containing:

(1) Date and time of accident etc.
(2) Name, occupation and nature of the injury to the person affected.
(3) Place of injury or where the dangerous occurrence happened.
(4) Brief description of the circumstances.

THE HEALTH AND SAFETY AT WORK ACT 1974

This Act made further provision for ensuring the health, safety and welfare of operatives at work, placing the responsibility on both employer and employee alike. Some of these are listed below.

Employers must:
(1) Provide and maintain safe plant and systems of work.
(2) Ensure that their activities do not endanger anyone.
(3) Provide information, instruction, training and supervision as necessary.
(4) Ensure a safe working place, including methods of access and egress.

Employees must:
(1) Take reasonable care of the health and safety of themselves and all other persons who may be affected by their acts or omissions.
(2) Co-operate with management in all health and safety matters.
(3) Not interfere with or misuse anything provided in the interests of health and safety etc.

SAFETY EQUIPMENT

Regulations apart it is surely commonsense to be sensibly and safely 'dressed' when working on a building site:

(1) Wear safety helmets at all times. A nut from a scaffold fitting will go straight through the skull from a height of 4½ metres!
(2) Wear protective footwear at all times. A nail through the foot, or even a crushed foot, is a common result of wearing trainers or similar inadequate footwear.
(3) Wear eye protectors in the form of goggles, safety spectacles, visors or face screens when using abrasive wheels/discs, saws, drills etc., and also when cutting bricks.
(4) Wear face masks when working in – or causing – a dusty atmosphere.
(5) Wear ear defenders when exposed to high noise levels.
(6) Wear suitable gloves when lifting or working with 'abrasive' materials; and also where the skin may come into contact with an irritant substance.

SOME FURTHER POINTS

- Every hour, someone in the building industry is either killed or seriously injured.
- Over 40 per cent of major accidents/injuries on sites involve falls from heights. Every year around 40 operatives lose their lives in this way.
- 20 per cent of all accidents in the building industry involve injuries sustained while manually lifting and handling materials and equipment.
- There are over 1000 eye injuries every working day, some resulting in total blindness.
- A large number of abrasive wheels/discs accidents occur, nearly 70 per cent of which are caused by using the wrong type of wheel/disc!
- Dermatitis is a non-infectious, inflammatory skin condition, normally caused when the skin comes into contact with an irritant substance such as cement or lime. This problem accounts for over half of all working days lost through industrial sickness.
- There are over 3000 reportable accidents each year of operatives being struck by falling or flying objects.
- In 1988 there were 157 deaths in the industry. Yet another increase!

PROTECTION OF EYES

The Protection of Eyes Regulations came into force

as long ago as 1975 and can be summarised as far as bricklayers are concerned by the following four points:

(1) The Regulations relate, among other situations, to "building operations and works of engineering construction."
(2) The employer must provide eye protectors to each person carrying out such work listed in 4 below, and every person so provided shall wear them while employed in that task.
(3) Persons provided with eye protectors have a duty to take care of them, not to misuse them and to report any loss or damage to them immediately to the employer for replacement.
(4) Eye protectors must be worn when "breaking, cutting, cutting into, dressing, carving or drilling by means of power or handtools, other than a trowel." In other words, students in workshops using lump hammers and bolster, brick hammers, scutches and the like *must* be provided with – and should wear – eye protectors.

FIRST AID

The Construction (Health and Welfare) Regulations 1966 and the Health and Safety (First Aid) Regulations which came into operation on 1 July 1982 state the requirements regarding the provision and contents of first aid boxes.

Where an employer has more than five persons in his employment he must provide a sufficient number of first aid boxes. Such boxes must be clearly marked 'FIRST AID' and be under the charge of a responsible person whose name must be prominently indicated either near or on the first aid box. Depending upon the number of persons employed on a site, the first aid box should be equipped as shown in table 1.1.

First aid is defined as "the skilled application of accepted principles given on the occurrence of any illness or injury." Courses to train operatives to become qualified first-aiders are held regularly throughout the country, since employers are required by law to have a trained first-aider on site, depending on the number of operatives on that site.

Table 1.1

Item	Numbers of employees				
	1-5	6-10	11-50	51-100	101-150
Guidance leaflet	1	1	1	1	1
Individually wrapped sterile adhesive dressings	20	20	40	40	40
Sterile eye pads with attachments	1	2	4	6	8
Sterile coverings for serious wounds (where applicable)	1	2	4	6	8
Safety pins	6	6	12	12	12
Medium sized sterile unmedicated dressings	3	6	8	10	12
Large sterile unmedicated dressings	1	2	4	6	10
Extra large sterile unmedicated dressings	1	2	4	6	8

Notes

1. The provision of triangular bandages and also sterile coverings for serious wounds is recommended. However, where the triangular bandage is sterile, this product satisfies both requirements.
2. Scissors are not to be kept in a first aid box. These should be kept in a locked drawer under the supervision of the first aider, along with any tablets such as aspirins, paracetamols etc.

The following basic principles of first aid should be the minimum common knowledge of all.

(1) Check for dangers, and ensure complete safety exists before proceeding.
(2) Ensure the casualty is breathing.
(3) Check for circulation. Is there a pulse?
(4) Stop severe bleeding; the loss of three pints of blood can result in loss of life.
(5) Never move a casualty with a back injury unless there are other dangers.
(6) With electrical injuries, turn off at plug or mains.
(7) Turn an unconscious or semi-conscious casualty into the recovery position (figure 1.1).
(8) Immobilise fractures in the position found. Do not attempt to straighten broken limbs.
(9) Reassure all casualties. Stay with them and send by-standers for help.

To conclude this chapter on safety, the authors feel the need to stress the importance in the correct use of plant, equipment, tools and apparel. All site personnel should ensure that all safety aids are used whenever there is the slightest possibility of an accident occurring.

When measuring the efficiency of a completed construction project, it is necessary to take into account the following:

(a) Did any accidents occur during the project?
(b) Were any injuries of a serious nature incurred?
(c) What were the total operative hours lost through accidents?
(d) Could the accidents have been avoided? Were adequate precautions taken? Has remedial action been taken?

No less important to safe work on a construction site is the protection of the public outside the site.

PROTECTION OF THE PUBLIC

Where the boundary of a site is adjacent to a public footpath or highway, a close-boarded fence must be erected to the satisfaction of the Local Authority, prior to the commencement of building operations. Under the Highways Act 1959, adequate protection of the general public must be provided in such a situation. The close-boarded fence, or hoarding as it is known, not only protects the public, but also the site and materials thereon, and defines the boundary for the contractor, ensuring for example that mechanical plant does not transgress beyond the work area.

A hoarding must be robust in its construction and capable of resisting impact damage and wind pressure, also acting as a shield between the site and the general public against dust and noise.

Before a hoarding can be erected in any public thoroughfare, the contractor must obtain a licence from the Local Authority who must be provided with full constructional details of the hoarding prior to issuing a licence, which usually takes between 10 and 20 days.

A hoarding is of course a temporary structure, consisting basically of posts, rails and sheeting of some kind (see figures 1.2 and 1.3). It should be not less than 2 m in height, and if it needs to be higher than this it may also require rakers and anchors (figure 1.4). Concrete posts (spurs) may be used for the initial supports, let well into the ground and held in position by a mass concrete surround. Timber posts are bolted to these spurs and connected by horizontal rails which in turn support the sheeting (figure 1.4). Scaffold tubing too is useful for hoardings, faced for example with corrugated sheeting. A hoarding should be painted white and bulkhead warning lights should be attached at intervals along its length.

Figure 1.1 The recovery position

It is important for obvious reasons that any doors in a hoarding are sliding or open inwards and that if the hoarding encroaches on the footpath the ends are splayed rather than square (figure 1.5). Furthermore, if a footpath is so restricted that pedestrians have insufficient space, further permission must be obtained to erect a temporary walkway in the road.

Where operatives are working on scaffolding, above hoardings, the scaffold may need to be completely enclosed in heavy gauge, polythene sheeting or similar; alternatively a fanguard may be considered

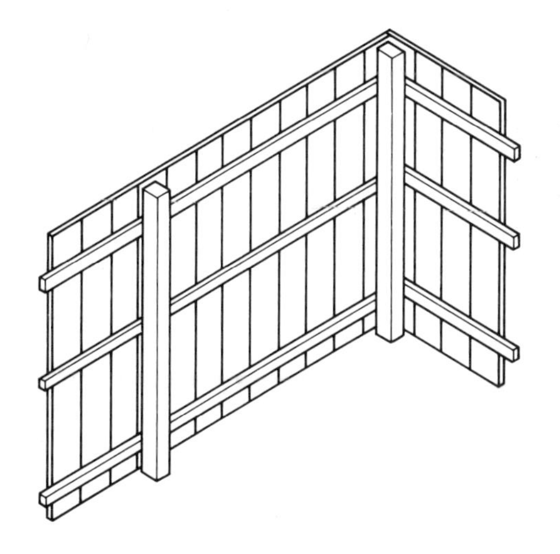

Figure 1.2 Internal elevation showing a return end of a hoarding consisting of posts, rails and boards

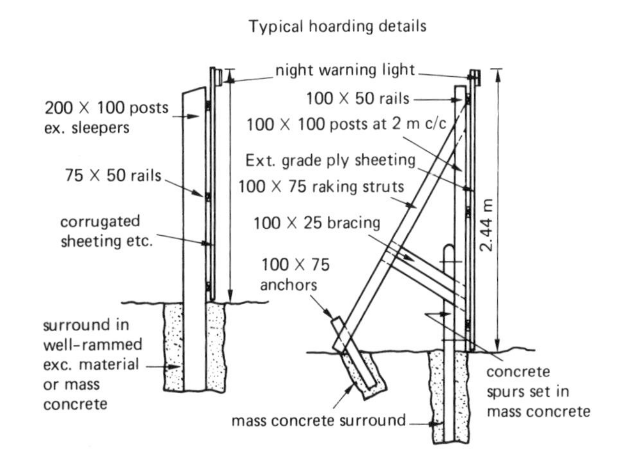

Figure 1.3 *Figure 1.4*

adequate to protect the public from falling objects (see figure 1.6).

The fan transom is fastened below the inside ledger and above the outside ledger, giving a minimum slope of 1 in 12. Both the platform and the fan must be fully boarded out whether or not this lift is a working platform. A fan transom must be situated in every bay.

The arrangement shown in figure 1.6 is described as light/medium duty, this being considered as a suitable form of protection against falling bricks and similar small objects. The strength of the fan transom may need to be supplemented by raking tubular struts supported by the ledger below if anything heavier than this is likely to fall (such as for demolition work etc).

Where fans extend over roadways, the minimum height should be 5 metres.

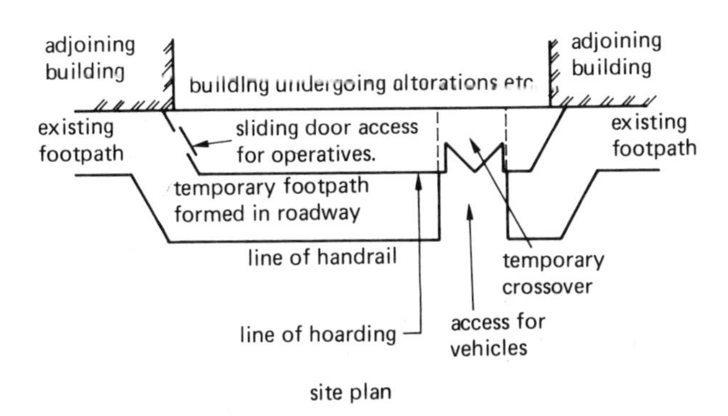

Figure 1.5 Where a hoarding encroaches on a footpath

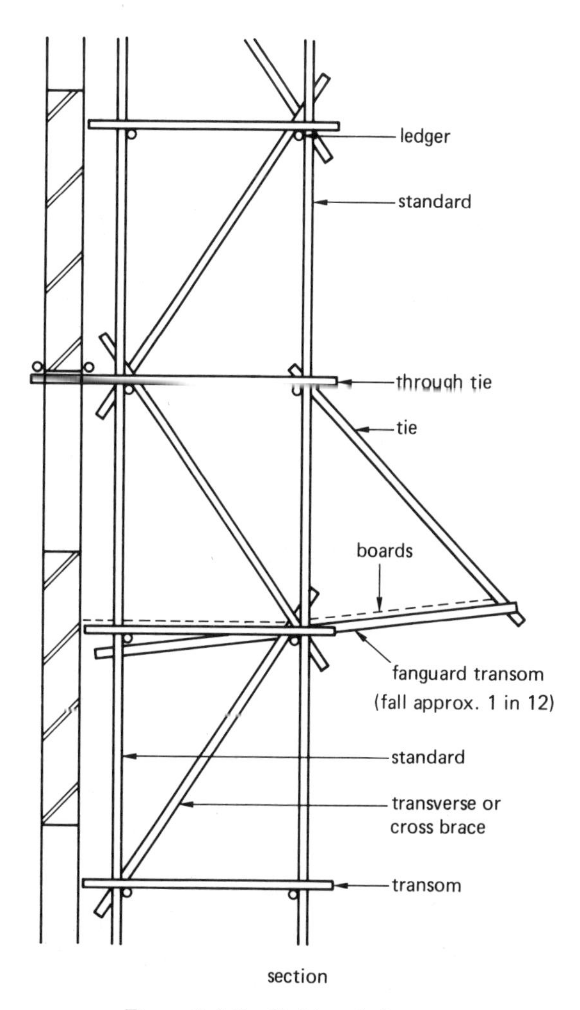

Figure 1.6 Scaffold with fanguard

2
DECORATIVE AND FUNCTIONAL FEATURES

In the construction industry designers are now taking advantage of the modern technology that enables the manufacturers to supply materials considerably superior in quality to those available years ago. The result is that many newly constructed buildings have façades consisting of contrasting-coloured bricks with surface textures chosen to suit the particular environment.

While colour and texture have improved we find that many buildings lack the decorative aesthetic qualities that are to be seen in older buildings; decorative face bonds are seldom used, and buildings now constructed with drab and plain face areas could be greatly enhanced by including decorative features in the brickwork.

STRING, DENTIL AND DOG-TOOTH COURSES

String Courses

These are sometimes known as band courses and are used to provide a distinct horizontal break in the façade of a building. They form subdivisions and interrupt the continuity of the facework, usually at storey height. A string course may be recessed, flush or projecting, and soldiers are often used for this purpose. Any decorative feature passing right round a building around storey height, possibly consisting of contrasting bricks or a different bonding arrangement, is called a string or band course. See figure 2.1*a*, *b*.

Dentil Courses

These are header courses, with alternate headers either projecting from the wall face or recessed. The projection or recess is usually kept to 19 mm, increasing to a maximum of 27 mm (figure 2.1*c*). Soldiers are occasionally used for this purpose (figure 2.1*d*).

In order to form a string or band course, dentils can be used between oversailing courses (figure 2.1*f*). When building projecting string courses such as these, the bricklayer's line must be fixed on the bottom edge rather than the upper edge in order to provide a perfectly straight line (figure 2.1*d*).

Dog-tooth Courses

Dog-tooth courses are another means of forming a decorative effect on the face of brick walls; the dog-toothing is obtained by setting each brick at 45° to the wall face. Dog-tooth courses may be projecting, flush or recessed from the wall face. Bricks are used flat or on end (figure 2.2); they should be set to line and checked for accuracy and position with a triangular templet (figure 2.3).

To provide a more decorative effect and increase the depth of band, continuous courses of dog-toothing are often used (figures 2.2 and 2.4). It is also common practice to form dog-tooth courses between the oversailing courses. When bricks are laid flat to form dog-tooth courses in walls one brick thick, half bats are used to form the dog-toothing, thus allowing for a fair face to be obtained on the opposite face of the wall. When the wall is over one brick in thickness, a stretcher course can be used at the back of the dog-tooth courses.

DECORATIVE TREATMENT OF QUOINS

The external quoins of a building are perhaps the most obvious position to create a decorative feature, yet in modern construction this is rarely seen to be done. To create a decorative effect with a quoin, irrespective of the bonding arrangement of the walling, does not greatly increase the cost of labour, and the appearance of a simple return quoin can be greatly enhanced by the formation of either:

(a) an indented or recessed quoin
(b) a rusticated quoin.

(1) Indented or Recessed Quoins

These are formed by recessing a number of brick courses at the quoin over a short length of walling.

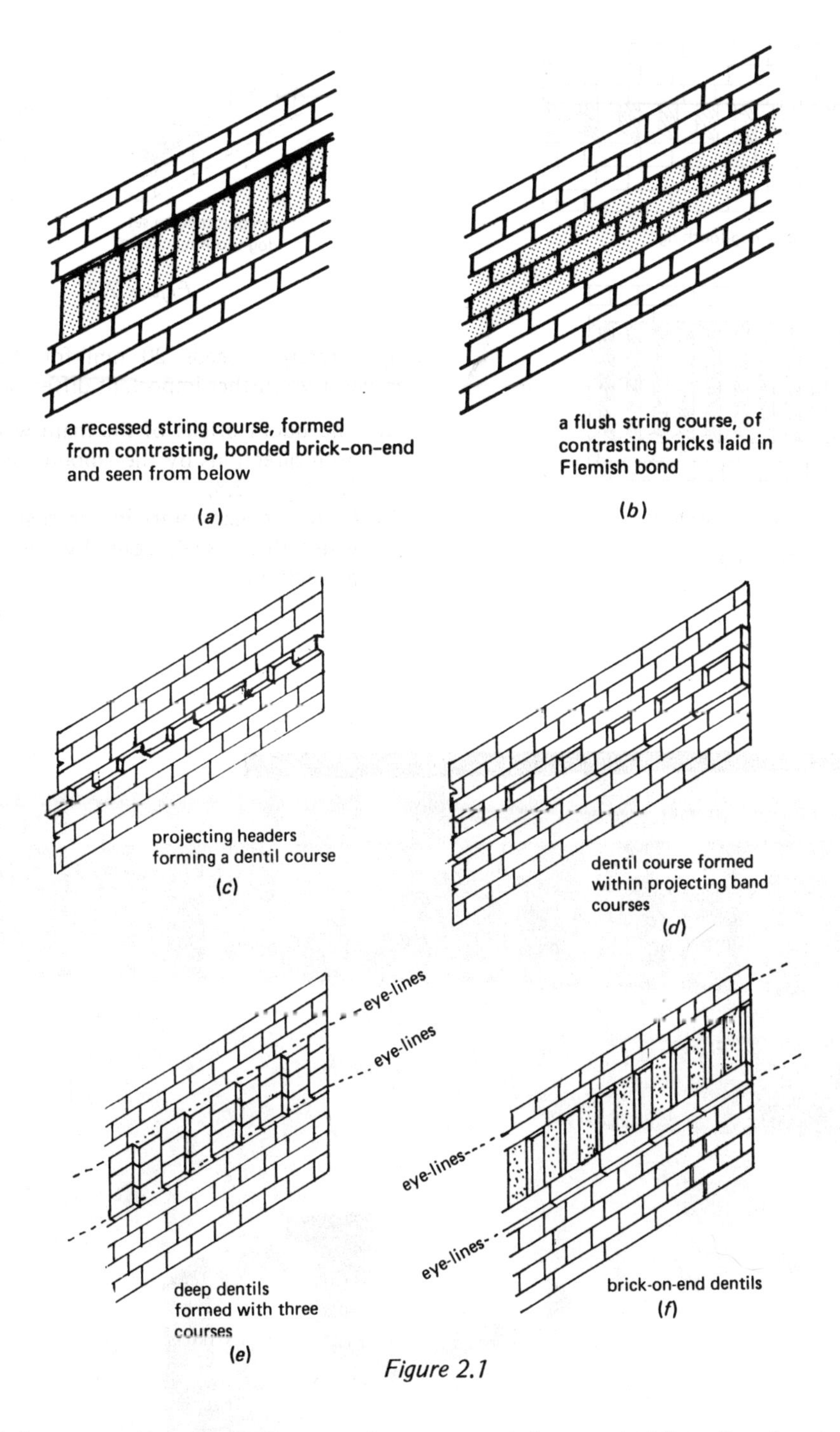

Figure 2.1

The height of the recessed work usually comprises one, two or three courses and the depth of the recess is rarely in excess of 20 mm since unless the bricks are very dense and weather-resistant any exposed arrises will tend to suffer from frost attack.

It will be obvious from figures 2.5 and 2.6 that the quoin bricks of the recessed work are cut on both header and stretcher faces in order to keep the perpends in a vertical line. Certain other recessed bricks, too, will need to be cut.

(2) Rusticated Quoins

These can basically be described as being the opposite to recessed quoins. Here the selected brick courses project over a short length and the projection once

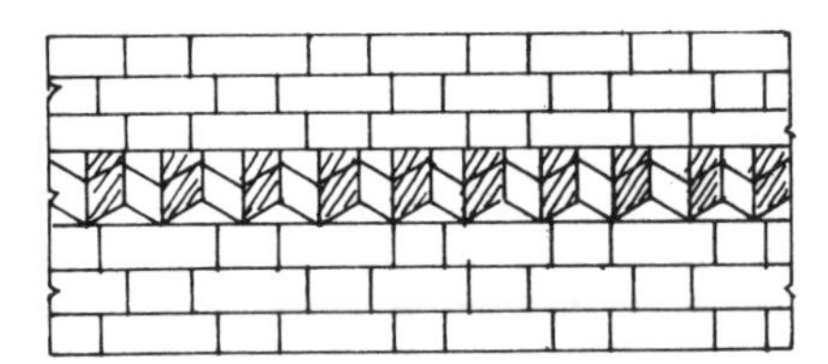
2-course dog-toothing

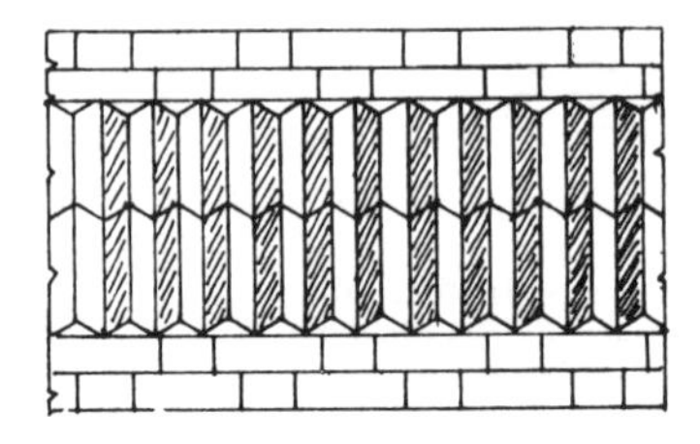
brick-on-end dog-toothing

Figure 2.2

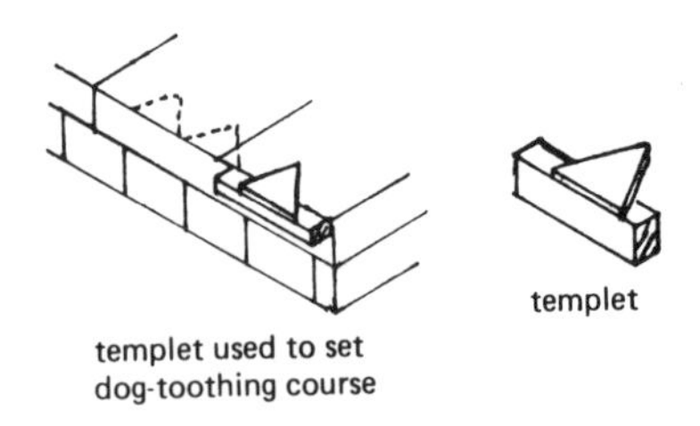

Figure 2.3

again rarely exceeds 20 mm for the reason given above. Two further important differences are:

(a) The quoin bricks of the main walling are cut in both directions by the amount of the projection (see figure 2.8).
(b) Certain collar joints in the rusticated work are wider than usual, again by the amount of the projection.

Figure 2.4

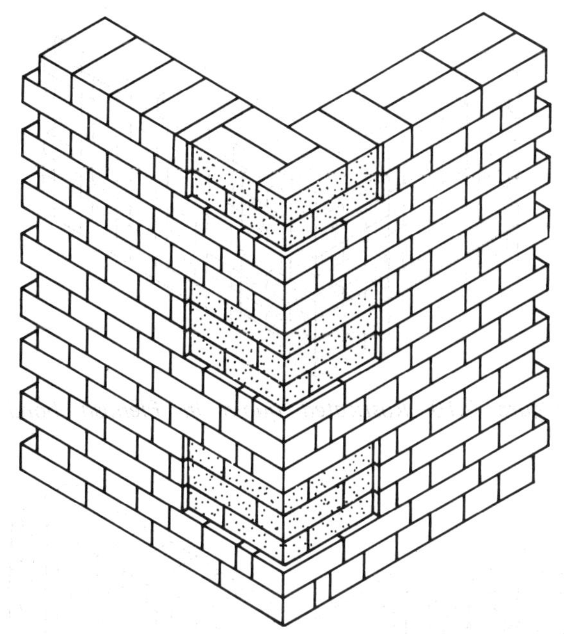

Figure 2.5 Indented or recessed quoin in English bond

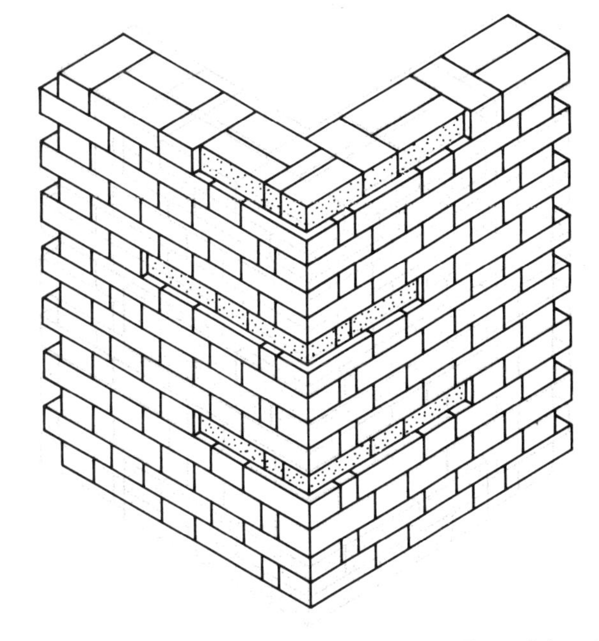

Figure 2.6 Indented or recessed quoin in Flemish bond

Figure 2.7

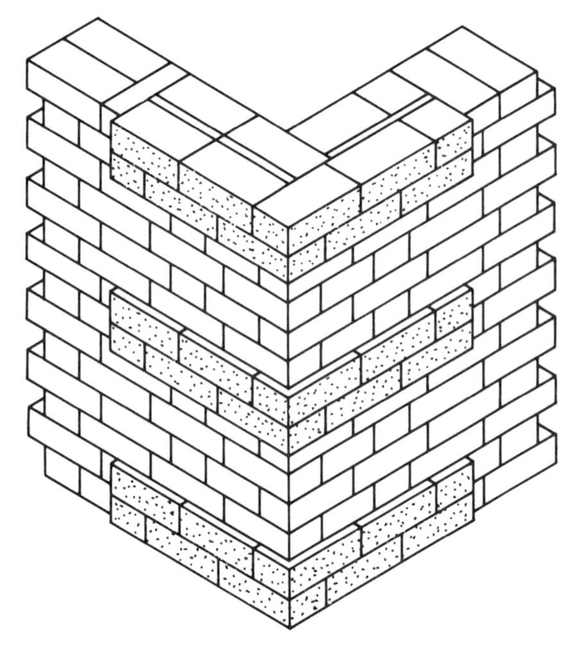

Figure 2.8 Rusticated quoin in Flemish Garden Walling Rustications in stretcher bond

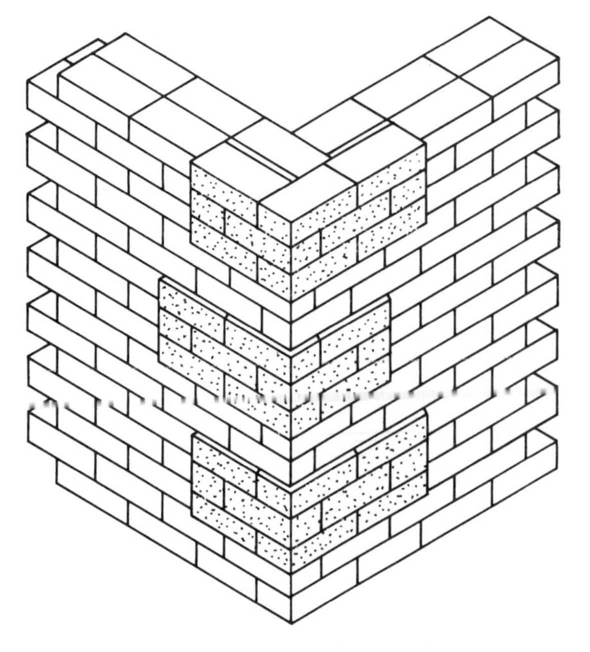

Figure 2.9 Rusticated quoin in stretcher bond with projections of alternating length

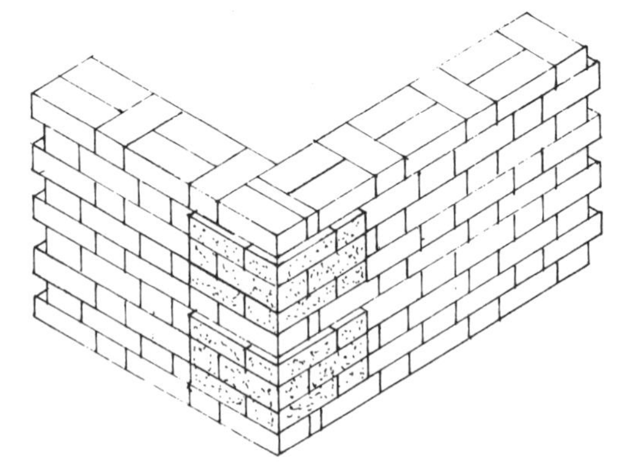

Figure 2.10 Rusticated quoin in Flemish bond

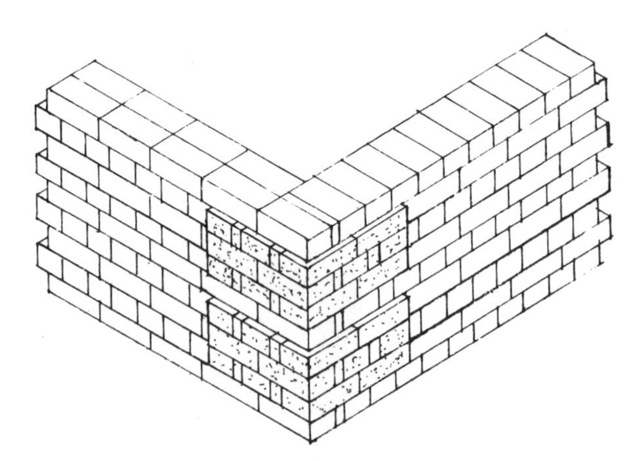

Figure 2.11 Rusticated quoin in English bond

PATTERN WALLING

On external walls the decorative effect can be enhanced by geometrical patterns formed on the face of the walling with contrasting-coloured bricks and projecting and recessed courses. When geometrical patterns formed with bricks projecting from the wall face are repeated either vertically or horizontally, the term used to describe the feature is strap work. Recessing fixed repeating patterns is termed coffering and it very often occurs when both of these decorative features are formed on the same wall face (figures 2.12 and 2.13).

Pierced Work

Pierced work is walling formed with repeating patterns, obtained by leaving voids or piercings in the walling. This type of wall is usually designed to form

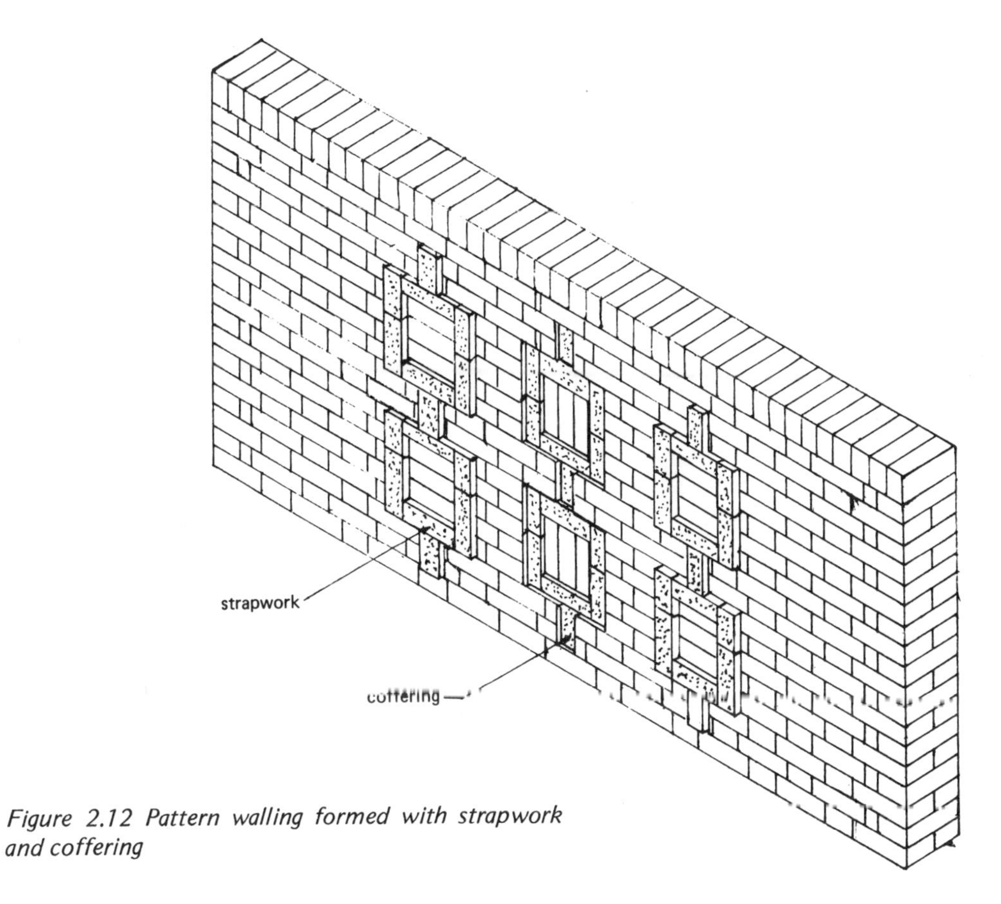

Figure 2.12 Pattern walling formed with strapwork and coffering

balustrading or screens. It is extremely effective and normally the wall is built with the same type of brick throughout (figures 2.14 and 2.15).

Decorative Internal Walls

Designers of modern brick buildings are now increasing the amount of exposed brickwork used internally, in both houses and other types of building. A study of internal decorative work shows that the use of geometrical patterns and light provides more aesthetic qualities for this type of walling than the use of contrasting-coloured bricks.

Figures 2.16 and 2.17 show the use of two types of zigzag bonding. The geometry of the design increases the depth of the decorative effect, while figure 2.18 shows the elevation of an internal wall built in herringbone bond which is formed to provide diagonal projections, even when built with the same type of brick throughout, as in figure 2.18. Light and shade provide the walling with a very interesting decorative effect, which increases with the dimensions of the walling.

Lap Work

Lap work is another form of bonding arrangement used for decorative internal walls. Again the same type of brick can be used for the entire wall face, or bricks of similar colour can be used. The designer will again obtain the decorative effect with patterns of light and shade (figure 2.19).

Concrete Screen Blocks

This type of precast concrete block is now becoming very popular as a method of forming screen or balustrade walls. The concrete blocks are made with fine aggregate and cement, formed in steel moulds and compacted by the vibration method. The dimensions of the concrete blocks vary from 300 x 300 mm to 450 x 450 mm, with a normal thickness of 75 mm. Geometrical patterns are formed within the area of each block and the blocks can be used effectively for wall heights of up to 2.4 m.

When the blocks are used for long lengths of walling it is advisable to provide end and intermediate supports in the form of attached piers, which are also

Figure 2.13

constructed of concrete blocks. These blocks are provided with a recess to accommodate the walling blocks. Plinths and copings can also be used to obtain increased stability and increase the decorative effect of the walling (figure 2.20).

DECORATIVE BRICK PANELS

The use of the panelled surface has long been a method of increasing the decorative qualities of walling. The use of the sunken or raised panel is often seen as a method of forming a feature in plain areas of walling and between piers. Placing of panels should be done with the utmost care. The dimensions and shape of the panel should coincide with the area of brickwork involved. Heavily sunken or too great projections often dilute the decorative effect intended for the feature.

A brick panel can be formed with a brick frame around the panel, or the panel can be formed within a recess in the walling. When a frame is used the base and sides should always be constructed before the panel is inserted, and should be formed with lines whenever possible (figure 2.21). If the panel is to project, the brickwork courses at the back of the recess are built using the bricks flat, but if a sunken or recessed panel is required, the courses of brickwork at the back of the recess should be formed with brick on edge and block indenting into the brick walling on each side.

Bond for Panels

Brick panels can be formed with basket-weave or herringbone bonds. Although diagonal and other bonding arrangements can be used, the designer usually favours the former to provide the decorative effect required (figure 2.22).

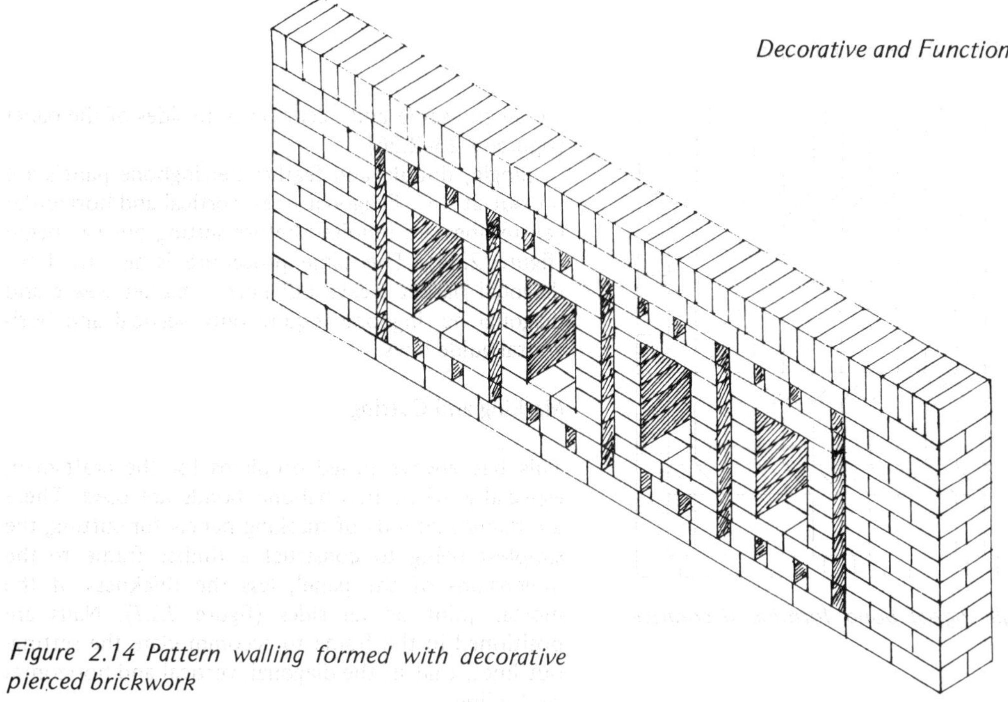

Figure 2.14 Pattern walling formed with decorative pierced brickwork

Figure 2.15

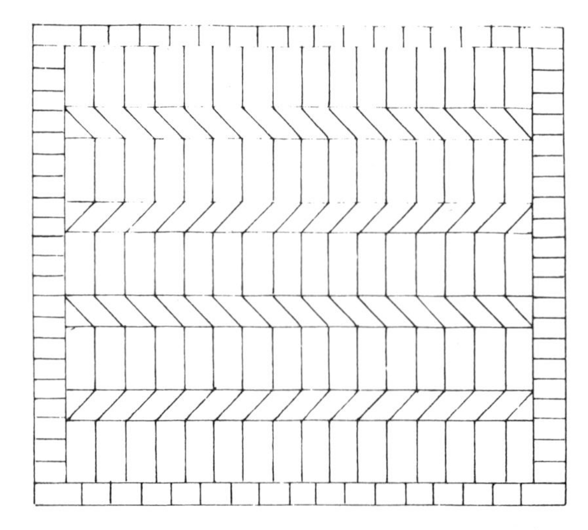

Figure 2.16 Single zigzag bond forming decorative panel (elevation)

Setting Out

Panels are usually set out according to

(1) the type of bond required
(2) the shape of the panel
(3) the dimensions of the panel.

When square panels are used, the setting out is always begun from the centre of the panel, whereas rectangular panels are always set out from the base line. This method is used to ensure that cuts, when required, are all the same and occur on both sides of the panel (figures 2.23-2.26).

Single, double and feather herringbone panels are set out at 45°. Diagonal lines, vertical and horizontal centre lines are required before setting out can begin (figure 2.24). The same procedure is also used for diagonal basket weave but normal basket weave and diagonal herringbone require only vertical and horizontal guide lines.

Marking and Cutting

This has always posed problems for the craftsman, especially when herringbone bonds are used. There are many methods of marking panels for cutting, the simplest being to construct a timber frame to the dimensions of the panel, less the thickness of the mortar joint on all sides (figure 2.27). Nails are positioned in the frame to accommodate the setting-out lines, that is, the diagonal, vertical and horizontal centre lines.

The brick panel is then laid out to bond on a flat level surface, with allowances for mortar joints between the bricks. The area should be considerably larger than the dimensions of the panel. The timber frame with the attached lines is then placed on the surface of the bricks, and can be positioned and adjusted to suit the geometrical setting out of the bond (figures 2.28 and 2.29). The perimeter of the panel can then be marked. Whenever necessary the frame can be repositioned and the panel checked for accuracy.

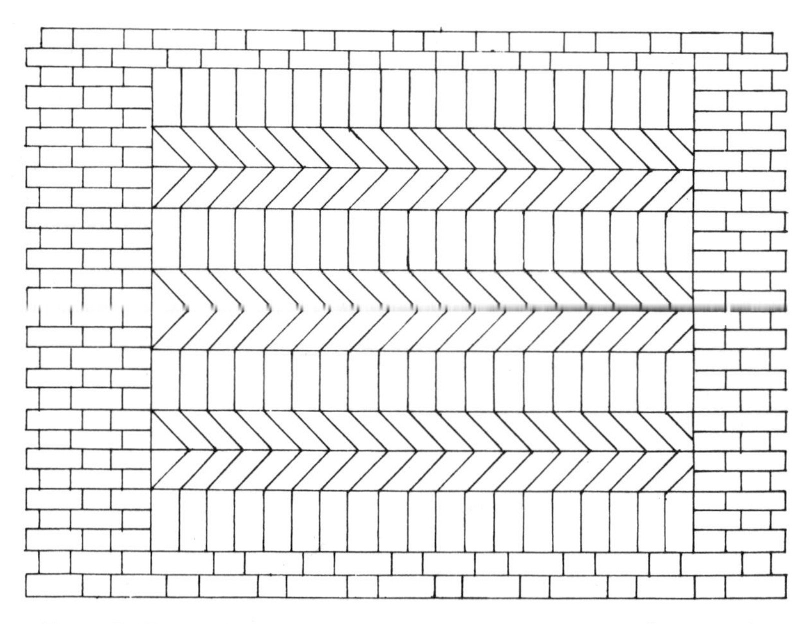

Figure 2.17 Double zigzag bond forming decorative panel (elevation)

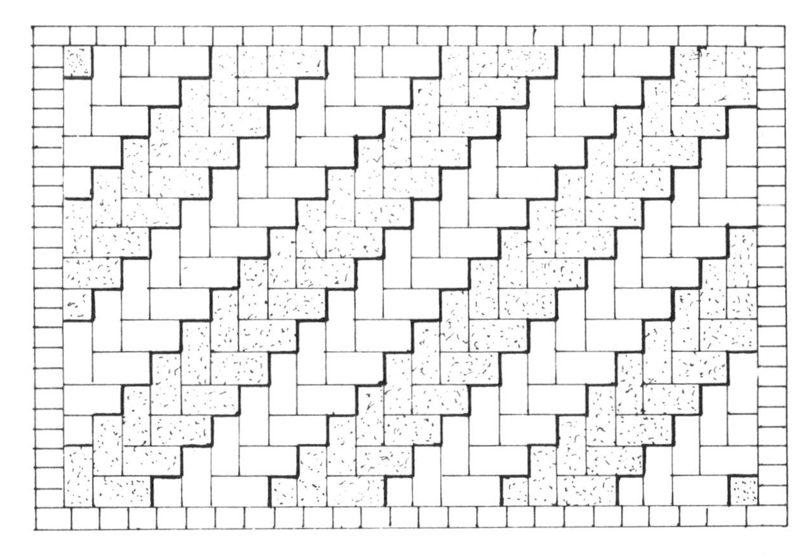

Figure 2.18 Internal decorative wall in diagonal herringbone bond using brick on edge courses and projecting contrasting bricks

Figure 2.19 Lap work

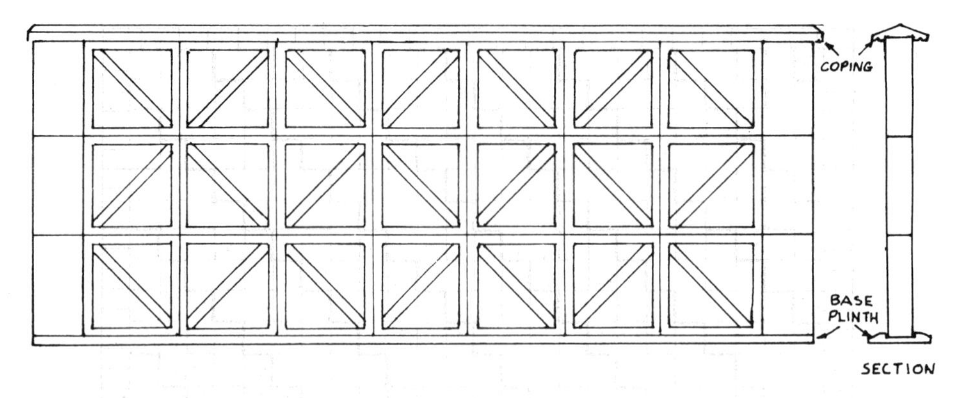

Figure 2.20 Screen or balustrade block wall with end supports

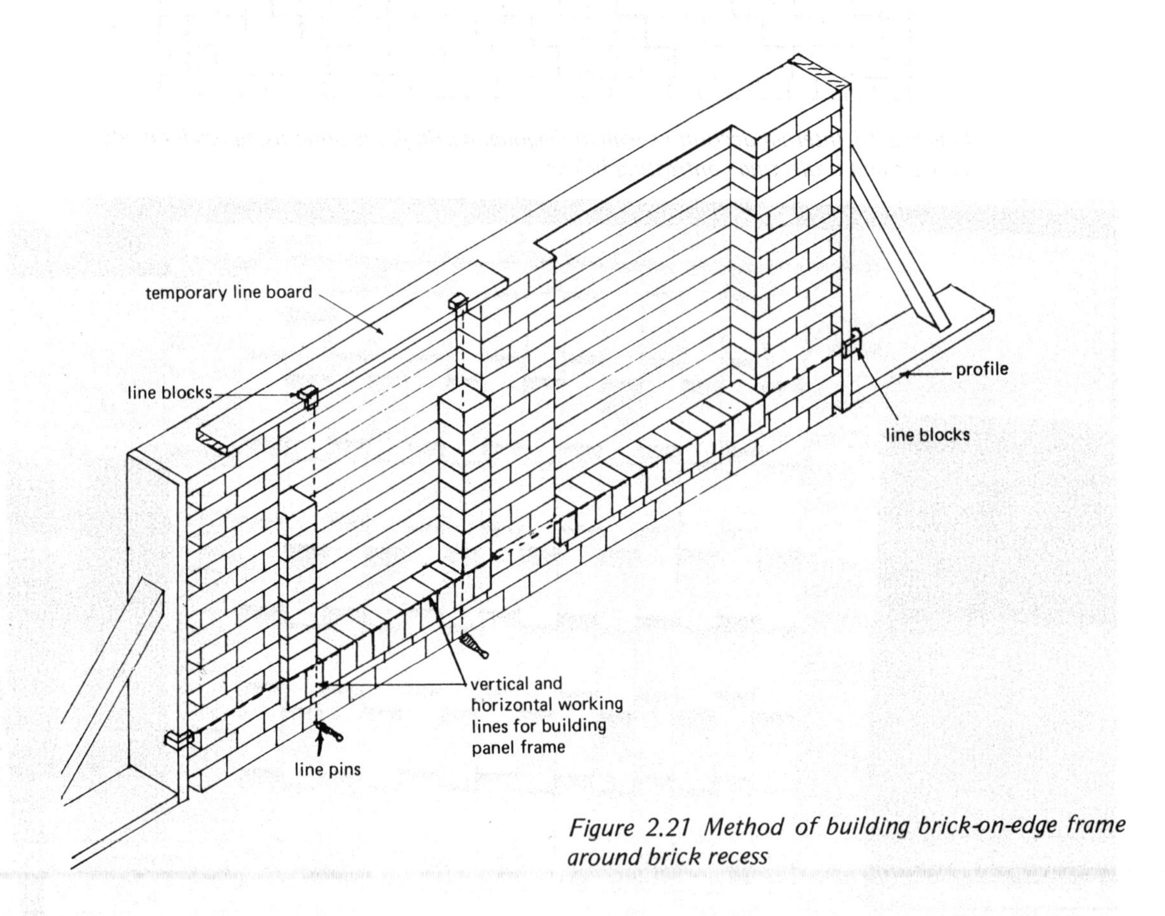

Figure 2.21 Method of building brick-on-edge frame around brick recess

Cutting the first three bricks for a single herringbone panel (marked 1 , 2 and 3 in figure 2.26) often causes problems, but these are quite straightforward. The complete method is explained in figure 2.30:

1 Stand two bricks on edge, face upwards, perfectly square to each other and leaving a 10 mm cross joint between the two.
2 Measure the exact length of one of the bricks, mark the same length from the corner onto the other brick and mark across in pencil.
3 Cut the bricks using a lump hammer and bolster. If an electric saw is used, the offcut will be too small.
4 The smallest offcut from the two bricks used will fit perfectly in the position shown.

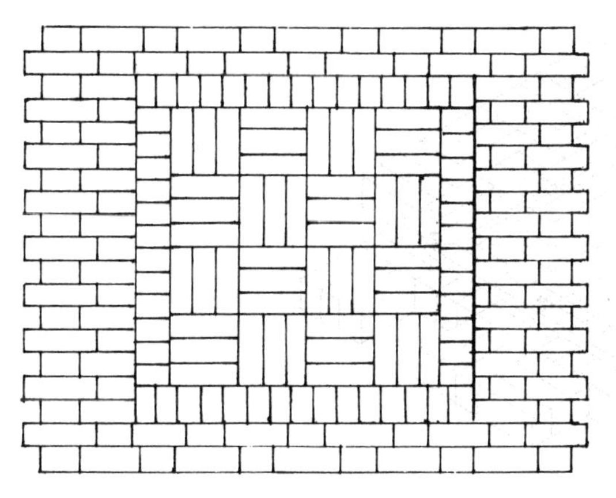

Figure 2.22 Basket-weave panel with brick-on-edge frame

Figure 2.23 Setting out a single herringbone rectangular panel

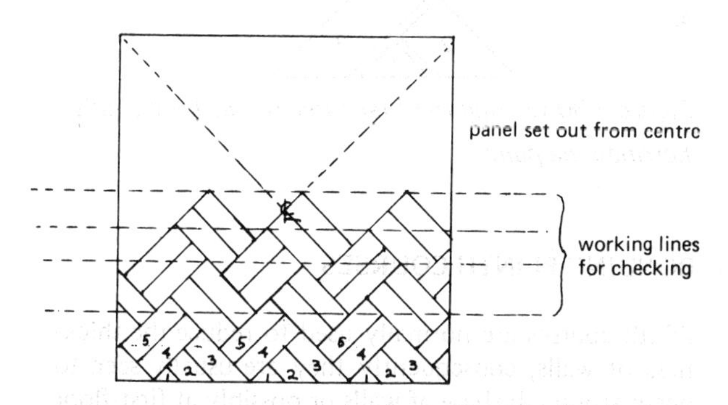

Figure 2.24 Setting out a double herringbone square panel

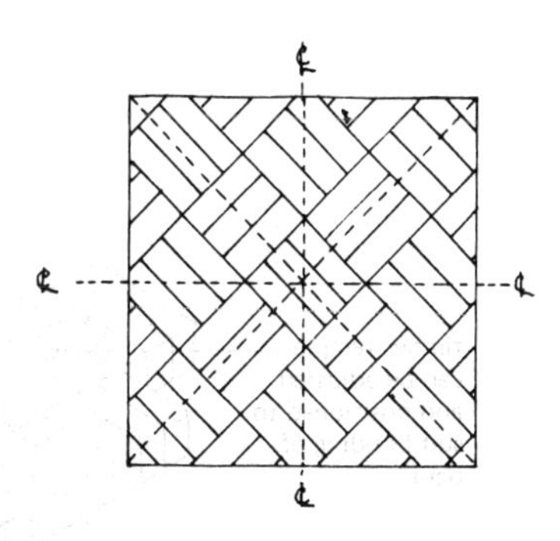

Figure 2.25 Setting out diagonal basket-weave in square panel

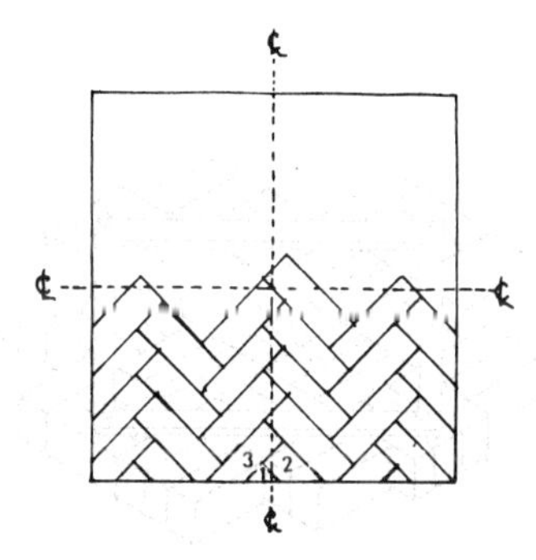

Figure 2.26 Setting out single herringbone in square panel

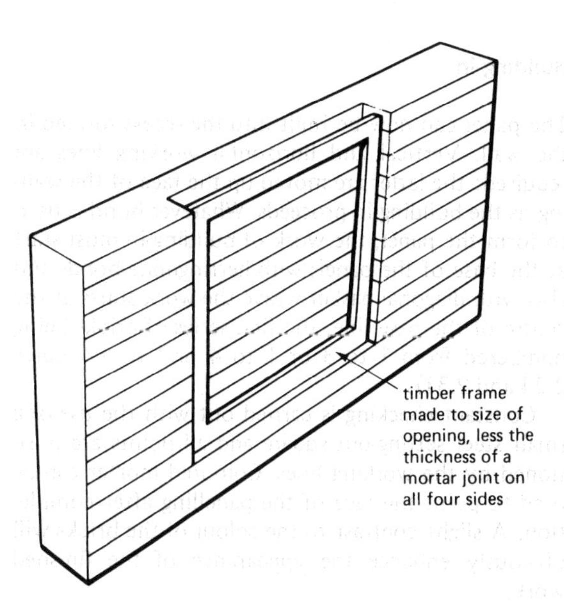

Figure 2.27 Fitting a frame to the panel recess

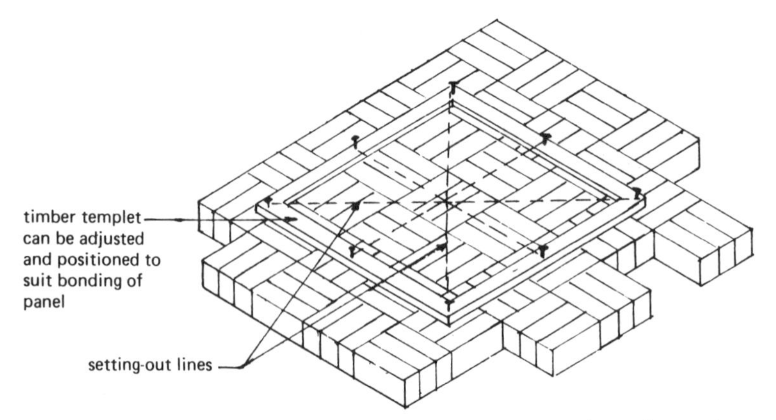

Figure 2.28 Use of timber templet for setting out and marking decorative panels

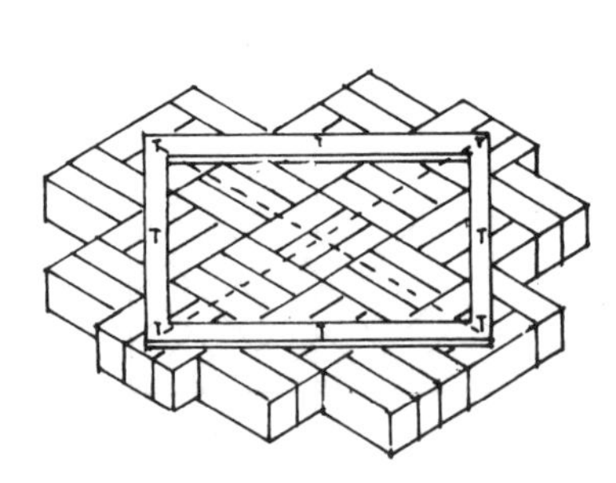

Figure 2.29 Setting out a diagonal basket-weave panel with a templet

Building-in

The panel can now be built into the recess formed in the wall. Vertical and horizontal working lines are required: the latter are moved up the face of the walling as the building-in proceeds. Whatever bond is used to form the panel, the work of building-in must start at the base of the panel; with herringbone bonds and also with diagonal basket weave the work starts at the centre of the base line, the first series of bricks being numbered from 1 to 3 or 1 to 4 or 1 to 5 (figures 2.24 and 2.31).

Constant checking is carried out with the use of a small steel setting-out square and all points are positioned by the working lines. Coloured mortar can be used to point the face of the panelling after completion. A slight contrast to the colour of the bricks will obviously enhance the appearance of the finished work.

Two examples of feather herringbone are shown in figure 2.32.

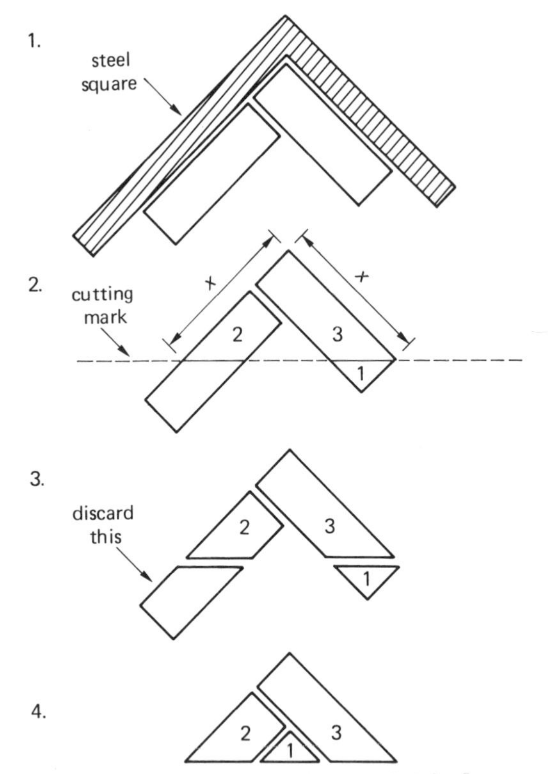

Figure 2.30 Cutting the first three bricks for a single herringbone panel

BONDING PLINTH COURSES

Plinth courses are normally used to reduce the thickness of walls, consequently they are usually seen to occur around the base of walls or possibly at first-floor level. The simplest form of plinth is an offset, where the bricks are set back from the face of the wall (figure 2.33*a*). With this method no special bricks are

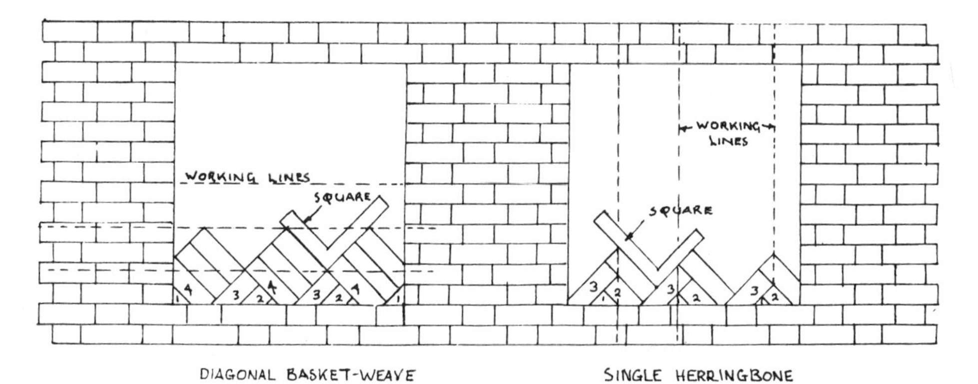

Figure 2.31 Method of building-in panels

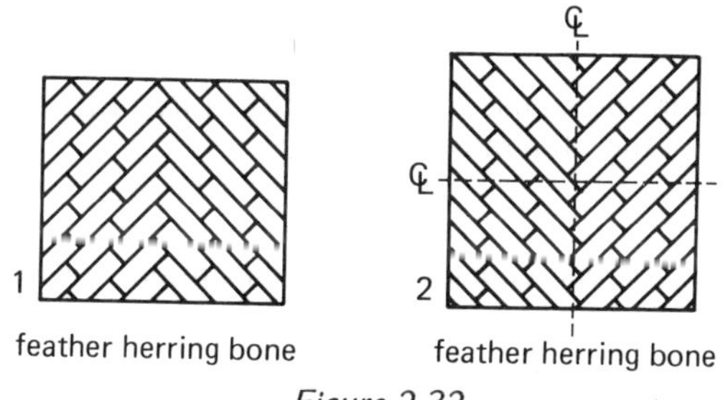

Figure 2.32

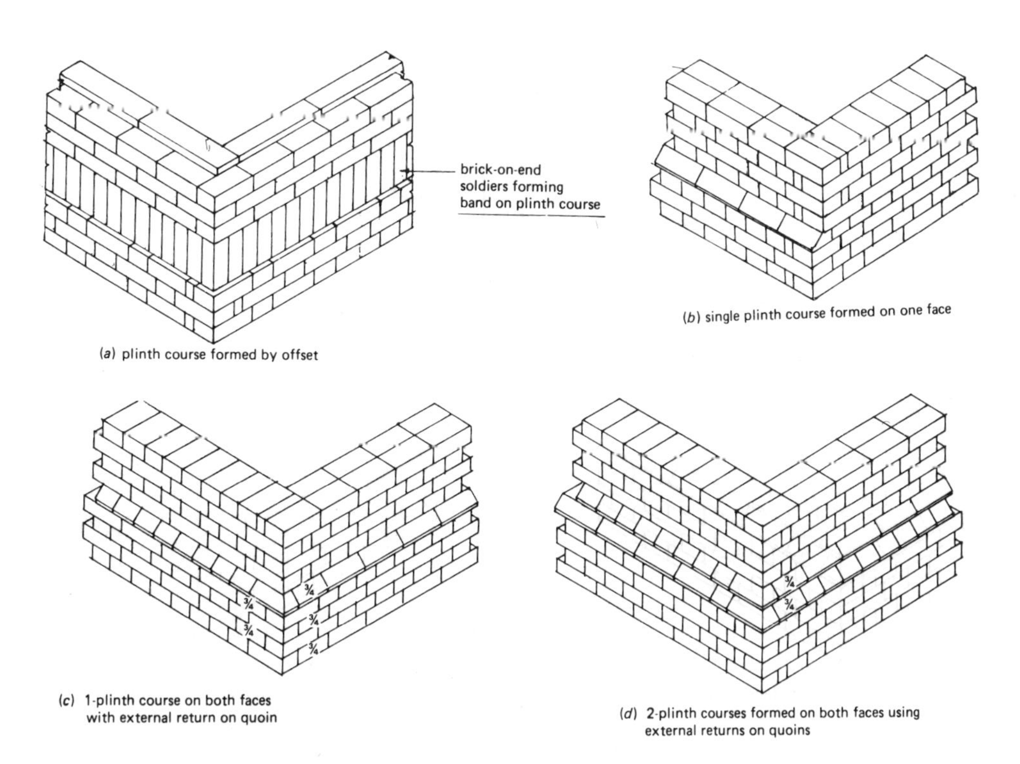

Figure 2.33 Plinth courses formed in English bond

required. If the wall is built in stretcher bond, the width of the cavity is increased below the offset course and the correct width of cavity is formed at the level of the plinth course. For solid walls the collar joint is increased in width to obtain the same effect.

The above method has its disadvantages because the offset formed provides a ledge that is always vulnerable to weather penetration even when a mortar fillet is applied. To provide a better form of weathering, plinth courses are normally formed with purpose-made splay bricks, which can be obtained in headers, stretchers and returns. Bonding these special plinth bricks and plinth courses has always posed considerable problems for the craftsman, but the problems can be greatly reduced by applying the following rules for bonding plinth courses.

Rules for Bonding Plinth Courses

(1) Always consider first, and bond in the course of brickwork immediately above the top-most plinth course.
(2) It is permissible to use a queen closer on the face of the wall, at a return quoin, for one course only; this brick is really one of two bevelled closers.
(3) It is acceptable to use header over header or stretcher over stretcher provided that there is a lap of 56 mm (figures 2.34*b* and 2.35).
(4) If necessary the plinth courses can consist of courses of stretchers, even if this is different from the walling bond.
(5) The courses of brickwork below the bottom plinth course must always be considered last and bonded accordingly. This may involve the use of broken bond, but this must be a secondary consideration, and the inclusion of cut bricks cannot always be avoided (figure 2.33*e* and 2.34*c*).

Figure 2.38 illustrates a decorative quoin constructed with plinth courses and inverted splay bricks.

CORBELLING BRICKWORK

It may be necessary during building operations to increase the thickness of walls, or form, or increase the dimensions of attached piers. The operation,

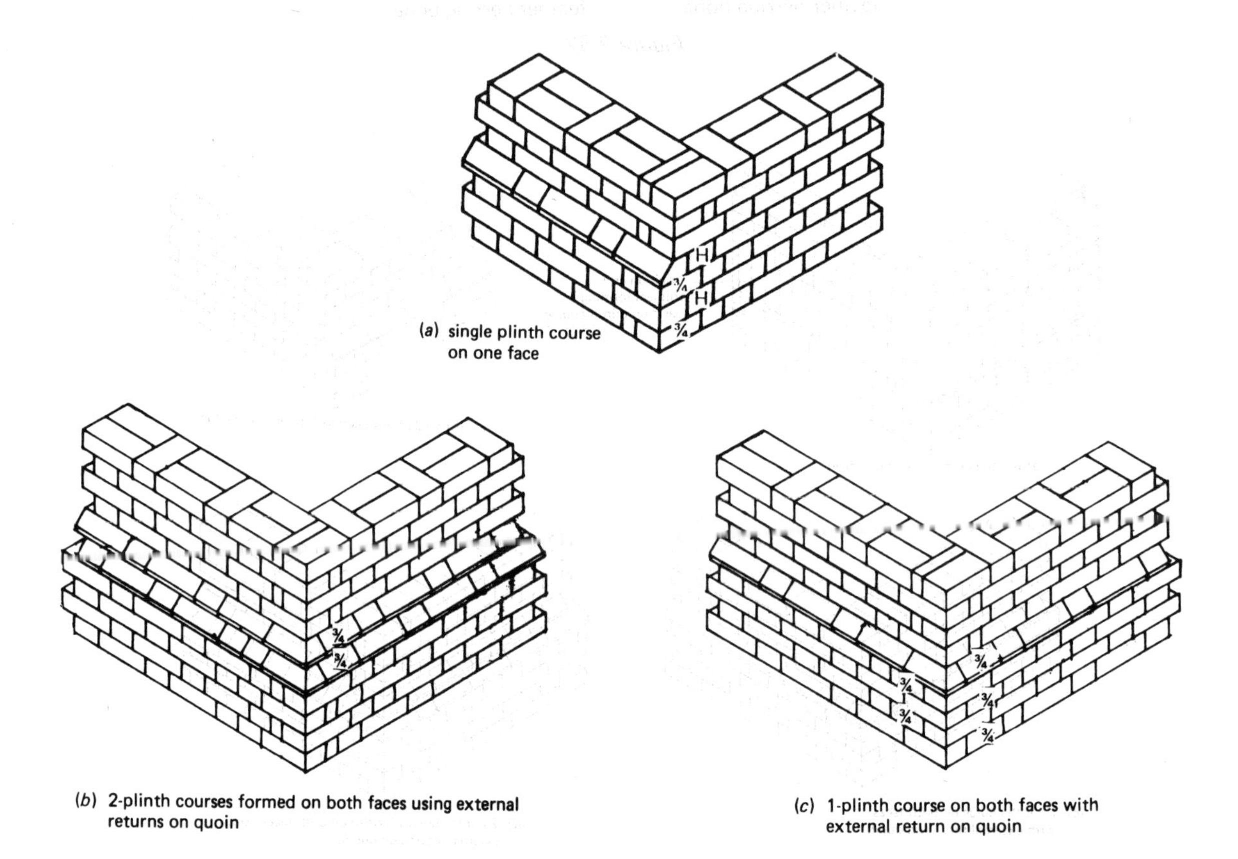

(*a*) single plinth course on one face

(*b*) 2-plinth courses formed on both faces using external returns on quoin

(*c*) 1-plinth course on both faces with external return on quoin

Figure 2.34 Plinth courses formed in Flemish bond

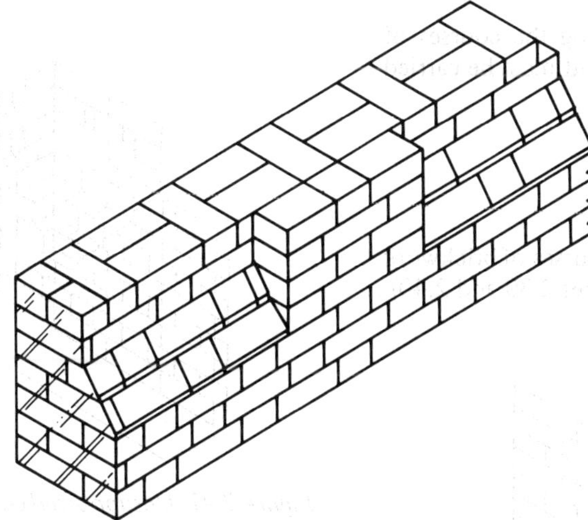

Figure 2.35 Plinth courses used to form an attached pier in Flemish bond

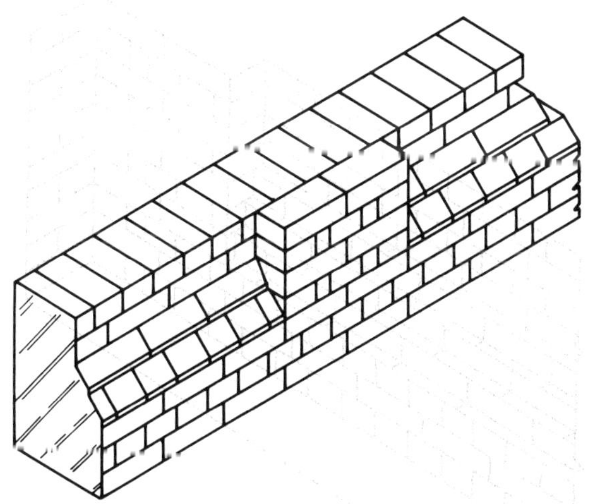

Figure 2.36 Plinth courses used to form an attached pier in English bond

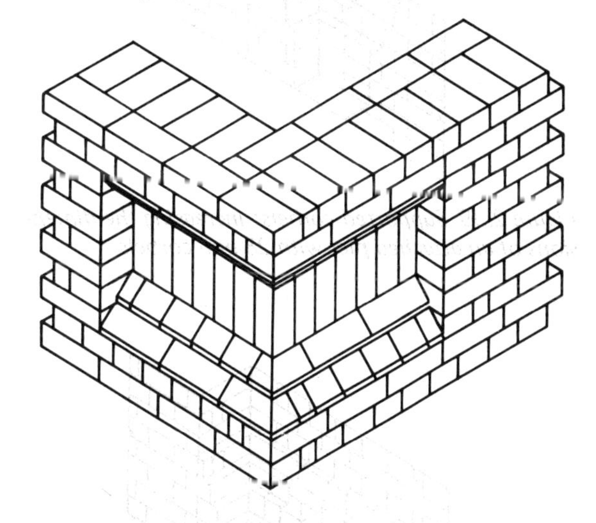

courses and inverted plinth bricks

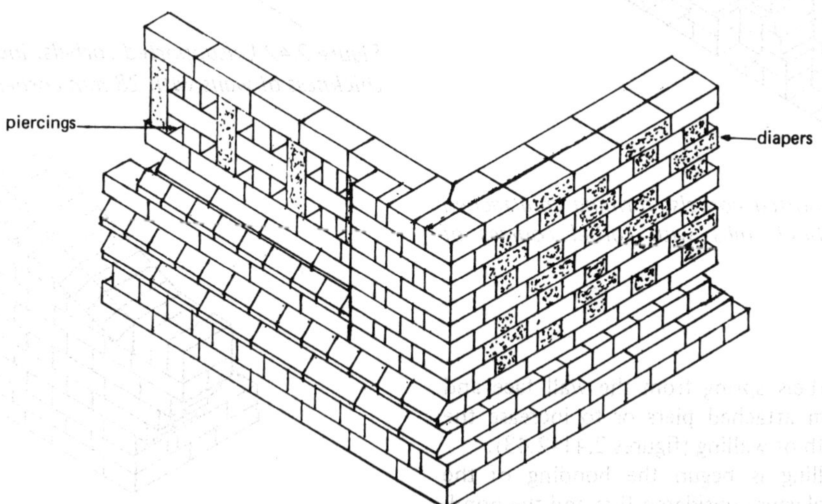

Figure 2.37 Method of forming plinth courses with splay bricks and offets, incorporating pierced walling and diapers

called *corbelling*, involves projecting the courses of brickwork beyond the wall face, and must be carried out with great care.

There are two types of corbel

(1) supported corbels
(2) unsupported corbels.

Supported corbels are projecting courses of brickwork that begin from attached piers (figures 2.39 and 2.40).

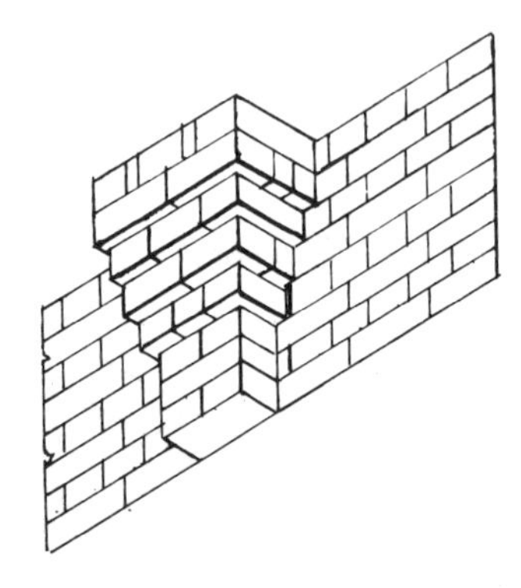

Figure 2.39 Supported corbels: increasing the dimensions of an attached pier with 28 mm corbels

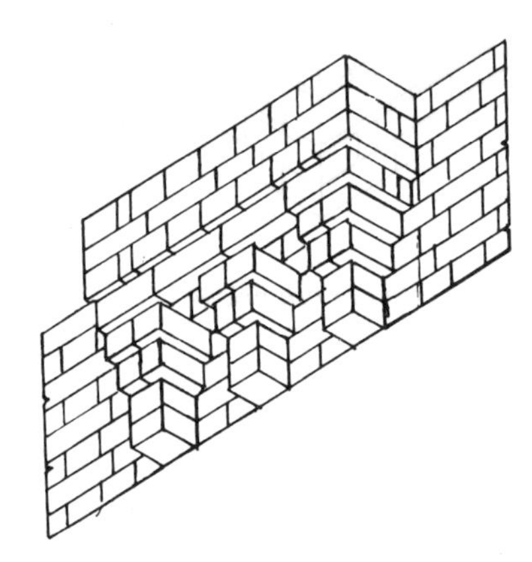

Figure 2.40 Supported corbels: forming an attached pier from half-brick nibs with corbel courses of 28 mm

Unsupported corbels spring from the wall face and are used to form attached piers or to increase the thickness or length of walling (figures 2.41–2.43).

Before corbelling is begun the bonding of the walling above is always considered first and the bonding for the corbelling course is then arranged to coincide with the work above.

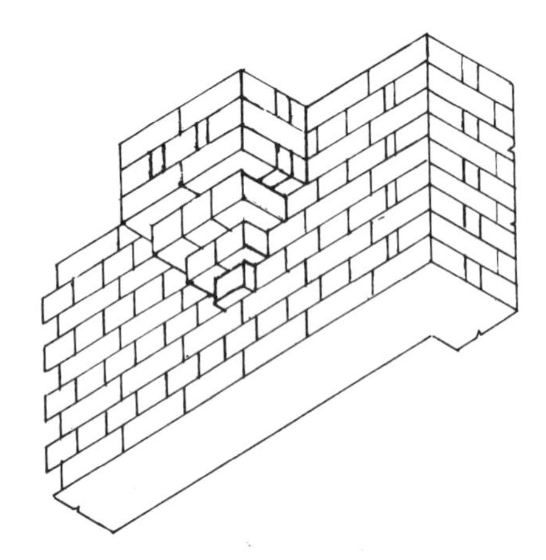

Figure 2.41 Unsupported corbels: forming an attached pier with 56 mm corbels

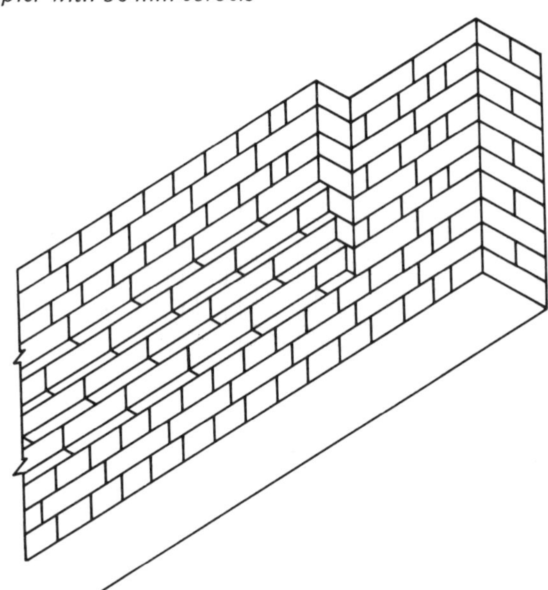

Figure 2.42 Unsupported corbels: increasing the thickness of walls with 28 mm corbel courses in

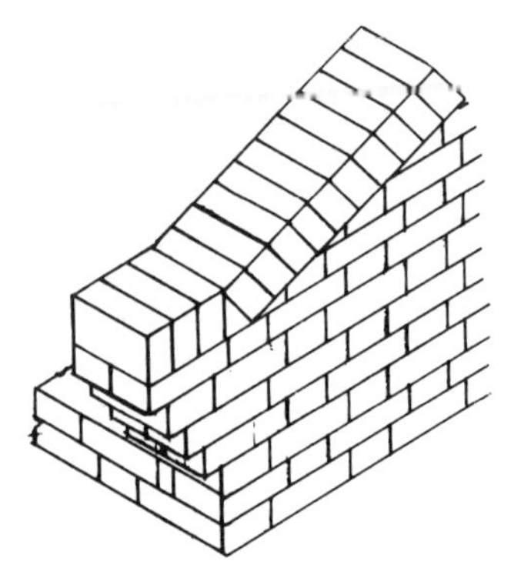

Figure 2.43 Unsupported corbels: a brick knee formed with 38 mm corbels

Rules for Corbelling

(1) All corbel courses should be arranged to project either 28, 38 or 56 mm. The maximum projection is 56 mm.
(2) Corbel courses should be formed in headers whenever possible.
(3) The amount of tie should be 168 mm whenever possible.

Rigid observance of the above rules is obviously not always possible, but tying into the existing wall *is* necessary to obtain stability. Therefore, backweight should always be provided on the tie courses before the next corbel course is begun.

The following recommendations should be observed for corbelling to be successful.

(1) Always use cement mortar for corbelling.
(2) Bricks used for corbels should be only slightly damp and should always be laid with frogs uppermost.
(3) The eye-line should always be formed along the bottom arris of the course (figure 2.44); individual corbels are fixed with the aid of a corbel templet (figure 2.45), while the corbel profile is used to check the completed projections (figure 2.46).

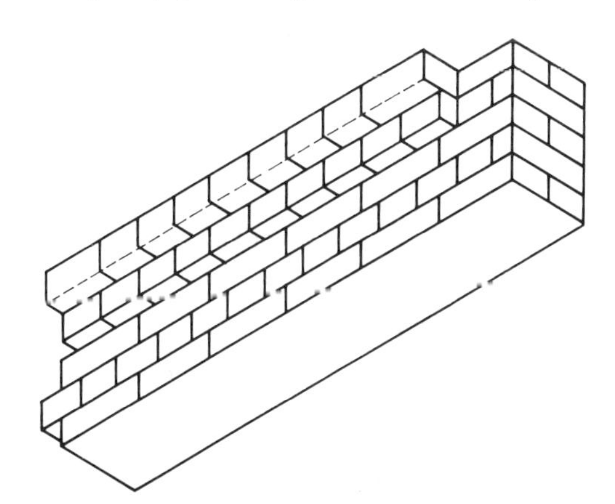

Figure 2.44 When building corbel courses the line is fixed to the bottom arris

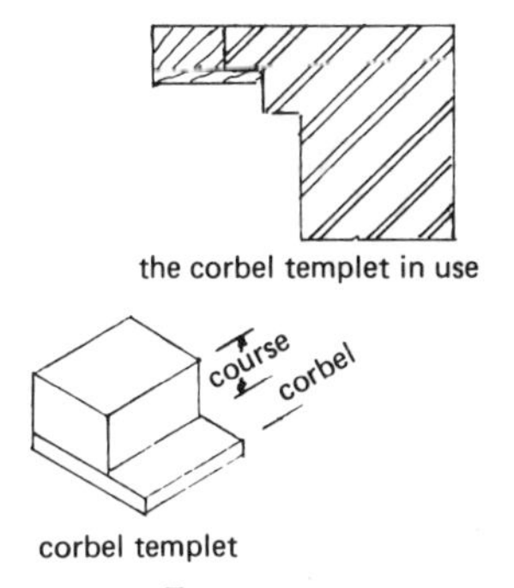

Figure 2.45

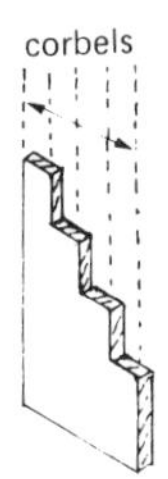

Figure 2.46 Corbel profile board for checking corbels

TUMBLING-IN

Attached piers and buttresses are often used to provide stability for walls that are subject to lateral pressure. These can be terminated with a simple concrete slab placed on top of the pier, or with courses of plinth bricks (figure 2.47), but where a brickwork finish is required, to provide strength and weathering and to be a decorative feature, the tumbling-in method should be used.

To construct tumbling-in brickwork correctly it is necessary to comply with the following rules.

(1) where the number of wall courses is odd, the number of tumbled courses should also be odd, but where the number of wall courses is even, the number of tumbled courses should also be even (figures 2.48 and 2.49); if one is odd and the other is even, this will result in broken bond which must always be placed BELOW the tumbling courses (see figure 2.50).
(2) the ratio of tumbled courses to horizontal buttress courses should always be 4:2 or 3:2 (figure 2.51)
(3) the tumbled courses should never enter the wall beyond half the thickness of the wall (figure 2.48)
(4) the bond for the buttress should be continued up the tumbled courses (figures 2.48, 2.50 and 2.52)
(5) a drip should be formed at the start of the tumbled-in work to allow water to fall clear of the buttress face (figure 2.48)

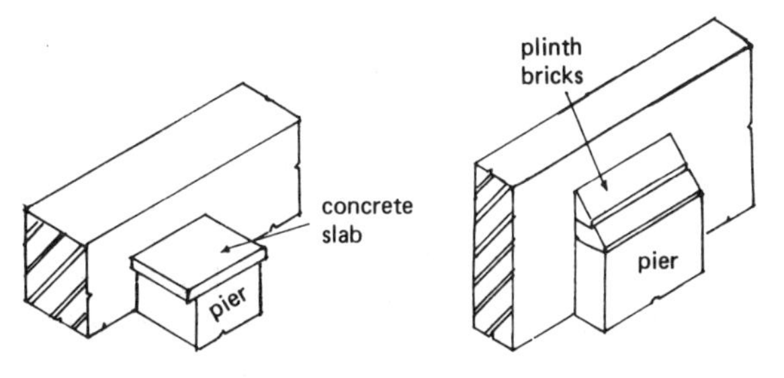

Figure 2.47 Methods of terminating attached piers

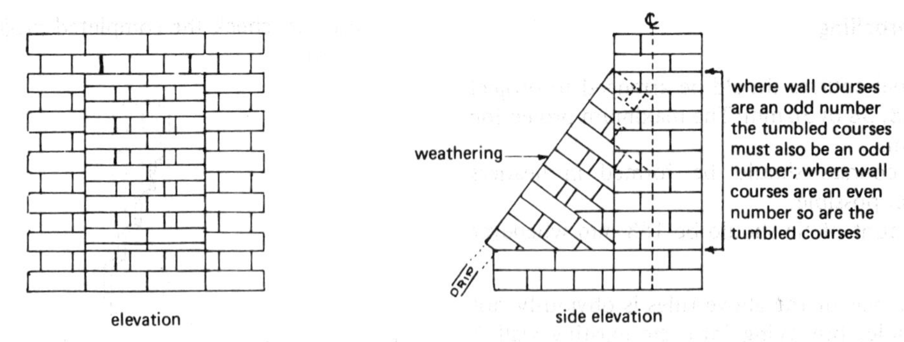

Figure 2.48

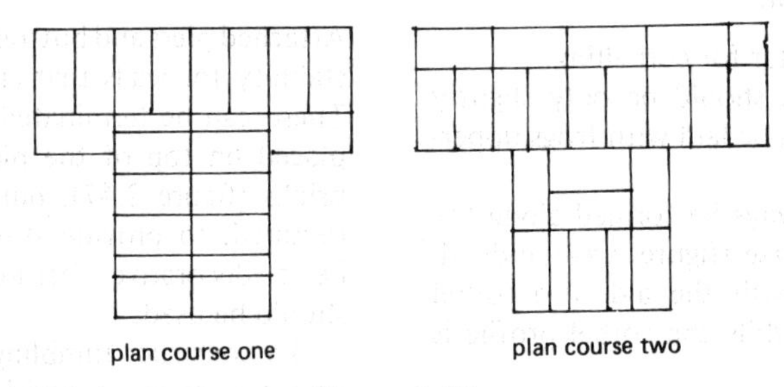

Figure 2.49

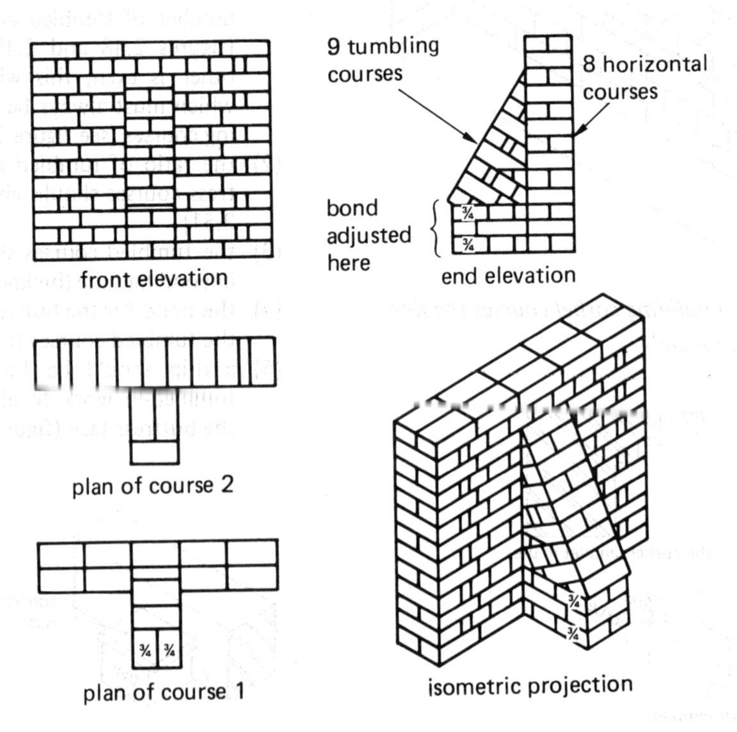

Figure 2.50

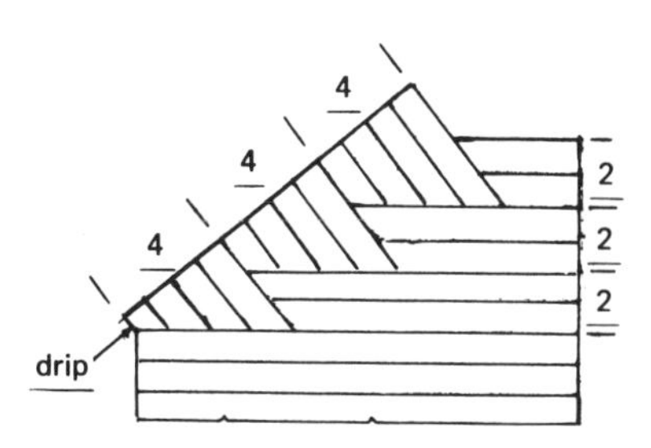

Figure 2.51 Demonstrating the rule that the ratio of tumbled courses to horizontal courses should always be 4:2 or 3:2

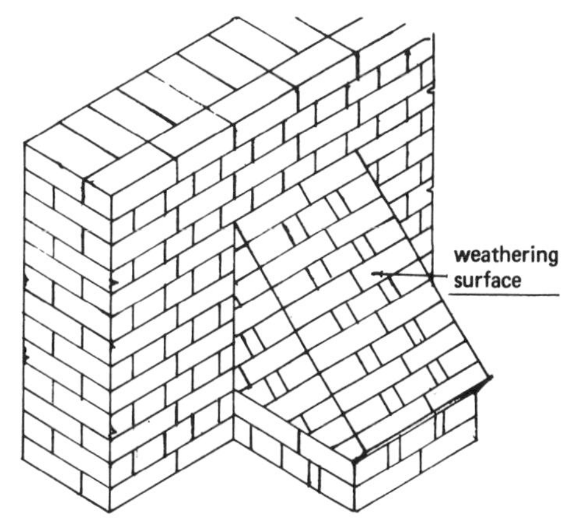

Figure 2.52 A tumbled-in cap for an attached pier, formed with inclined courses only

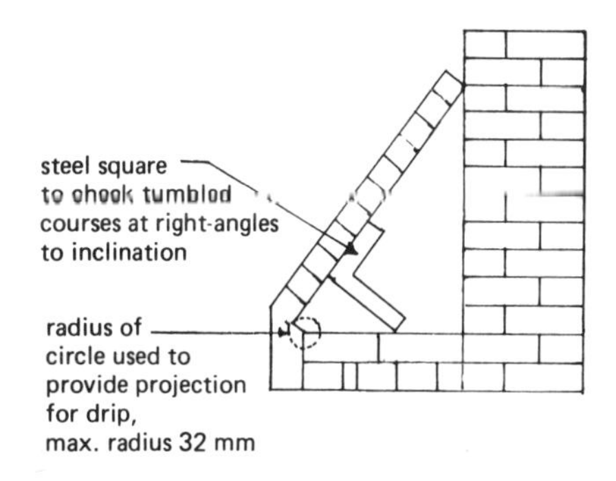

Figure 2.53 Building tumbled-in work: method of setting out the tumbling courses

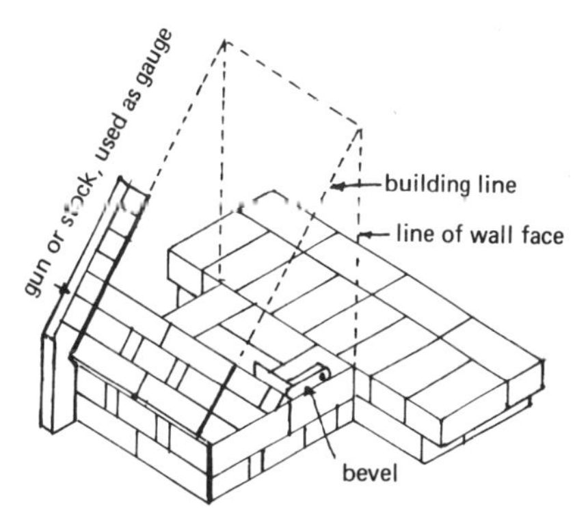

Figure 2.54 Building tumbled-in work: use of gun or stock to assist in the aligning of tumbled courses, with the bevel providing the angle of cut for the tumbled courses

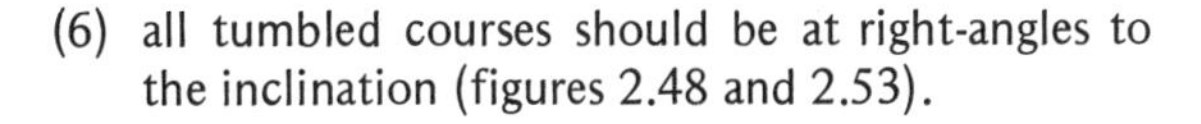

(6) all tumbled courses should be at right-angles to the inclination (figures 2.48 and 2.53).

To build tumbled-in brickwork it is necessary to erect building lines to provide the lines for the inclination; a gun or stock should also be formed, which can be used to check the surface of the work and also as a gauge rod for the tumbled courses (figures 2.53 and 2.54). To comply with rule 1 it may be necessary to adjust the courses of the tumbled-in work, by either increasing or reducing the bed joints.

Checking the tumbled-in courses during erection is carried out with a steel square to ensure that the courses are all at right-angles to the inclination (figure 2.53). The angle formed between the tumbled and the horizontal courses is obtained with a bevel, which can then be used to mark the angle of cut (figure 2.54).

To ensure that good effective weathering is provided it is essential that all tumbling-in courses are completely parallel to the angle of inclination, otherwise water may rest on any ledges that are formed. Cement mortar should be used and the bricks should be capable of resisting weather penetration. When tumbling-in work is accurately carried out, the buttress will always be a decorative and functional feature that demonstrates the skills of the bricklayer craftsman.

3
REINFORCED BRICKWORK

The designers of modern buildings are constantly using new techniques, methods and materials to provide structures with increased strength and reductions in loading. Because of this trend the use of reinforcement in walling is becoming a common practice. The reinforcement of brick walls allows for a reduction in wall thickness, and, when walls are reinforced above openings, the compressive strength is increased because the brickwork acts as a beam.

Foundation walling is often reinforced horizontally to prevent settlement, and vertically to resist lateral pressure (figure 3.1). Walls of concrete and brickwork in compound form are constantly used to provide strength and for their decorative qualities. This type of work is often carried out with the inclusion of rod and wire reinforcement (figures 3.6, 3.7 and 3.9).

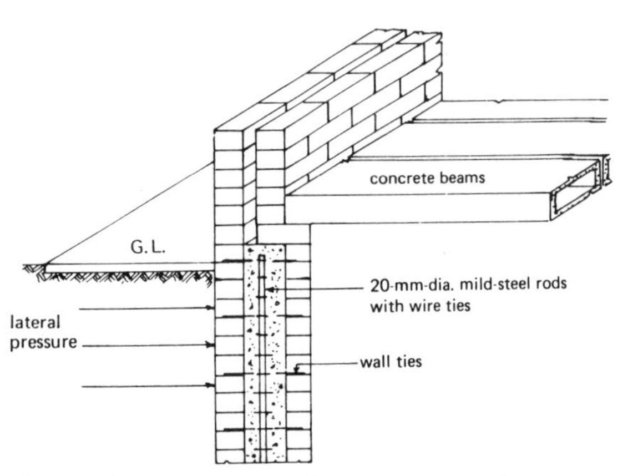

Figure 3.1 Vertical reinforcement in sub-structure

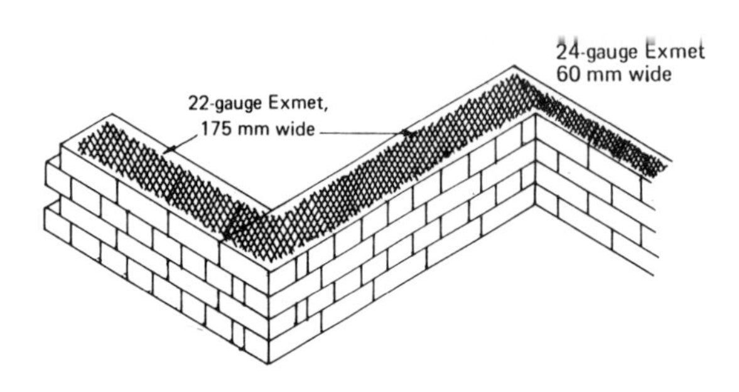

Figure 3.2 Reinforcing short returns using Exmet walling

TYPES OF REINFORCEMENT

Exmet Expanded Metal Mesh

This is a diamond-shaped mesh, obtainable in rolls of 18 m and in widths of 56, 175 and 300 mm. The gauge is 20, 22 or 24. To be effective, the mesh should be completely enveloped within the mortar bed. This type of reinforcement is used in walls and partitions to resist both horizontal and vertical pressure (figure 3.2).

B.R.C. Brickforce (figure 3.3)

This is a welded product consisting of two parallel wires joined by cross wires every 304 mm, all wires being approximately 3 mm thick. It is supplied in a variety of widths from 60 mm for half-brick walls to 160 mm for one brick walls, and in 3 metre lengths for ease of handling. Brickforce should be bedded and surrounded in mortar between courses of bricks or blocks and a wall reinforced with this material will have greatly increased resistance to horizontal pressures occurring from either side. It is also useful over openings, adding considerable tensile strength to the brickwork (figures 3.6 and 3.17).

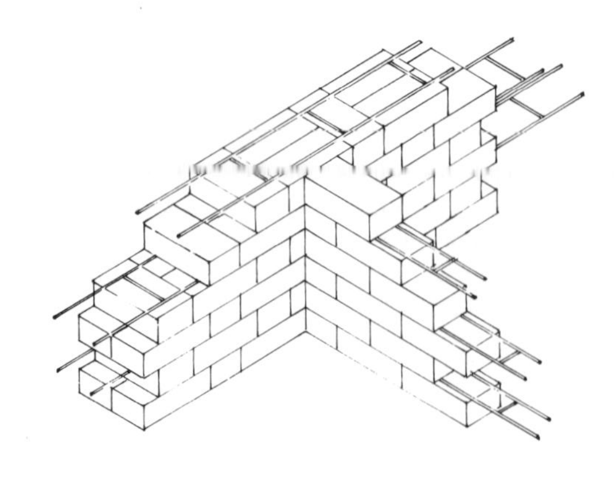

Figure 3.3 Horizontal reinforcement using Brickforce

B.R.C. Wallforce

This is similar in appearance to Brickforce but used to strengthen cavity walls. It is supplied in strips 3150 mm long, the width of each strip being 222 mm. Four main wires run parallel along the length of the strip, joined by welded cross wires at 450 mm centres, which take the place of conventional wall ties in cavity walls (figures 3.4 and 3.5).

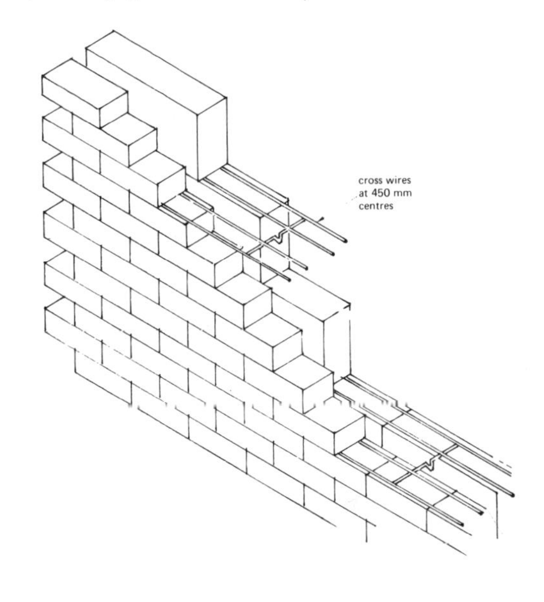

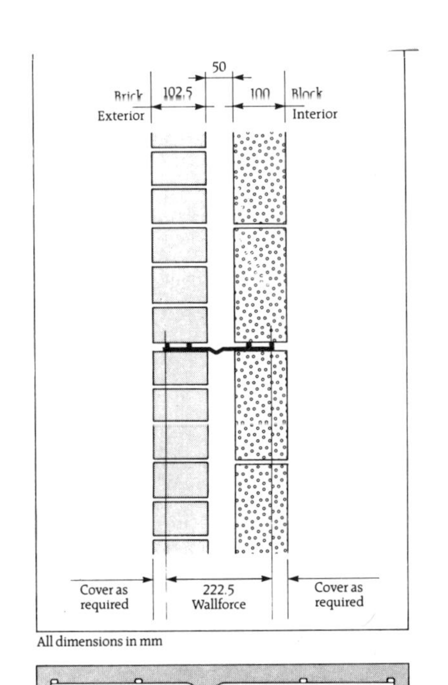

Figure 3.5 Wallforce

B.R.C. Bricktor (figure 3.7)

This is a stainless steel or galvanised wire mesh supplied in 25 m or 75 m rolls, in widths ranging from 50 to 300 mm. Some typical uses of Bricktor are:

(1) Carrying brickwork over door openings. By incorporating the mesh in the bed joints, conventional lintels can be omitted, often to the benefit of the appearance (figure 3.6).
(2) Bonding and strengthening corners and intersections of walls.
(3) Tying of walls to reinforced concrete frames.
(4) Resisting cracking of walls in areas of uneven settlement.
(5) General control of shrinkage and thermal cracking.

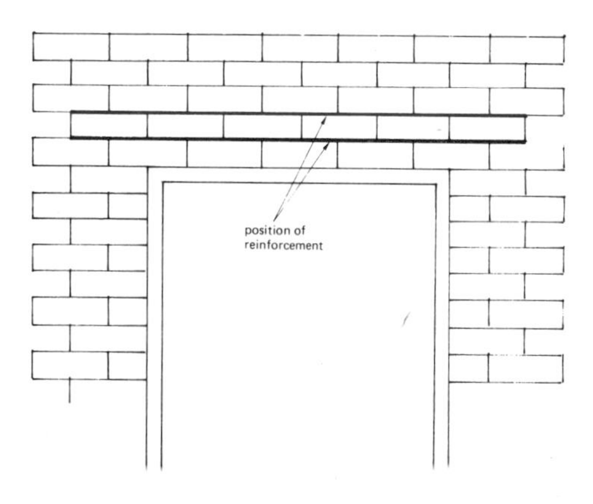

Figure 3.6 Using Bricktor over door openings

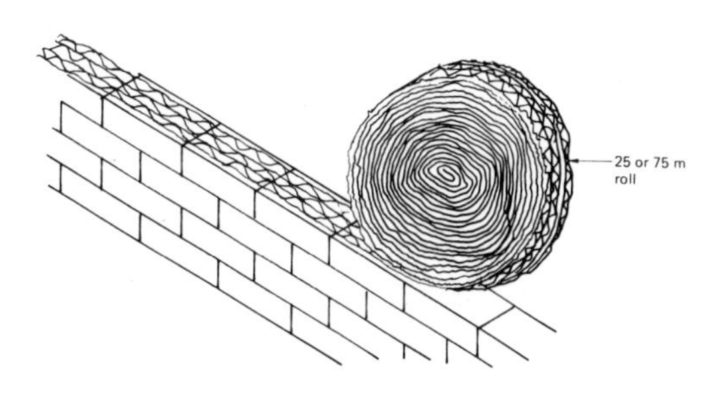

Figure 3.7 The use of Bricktor

Hoop-iron Reinforcement

This is a traditional type of horizontal reinforcement that is now only rarely used. The hoop-iron is 25 mm wide and 2 mm thick. It is possible to obtain it in galvanised form or with a bituminous coating. When used in brickwork it should be fixed within the mortar bed, with one band of hoop-iron for each half-brick thickness of walling (figure 3.8). As with other forms of horizontal reinforcement, the hoop-iron should be set at least 25 mm from the face of the walling. Joints are formed with welts or hooks to ensure continuity.

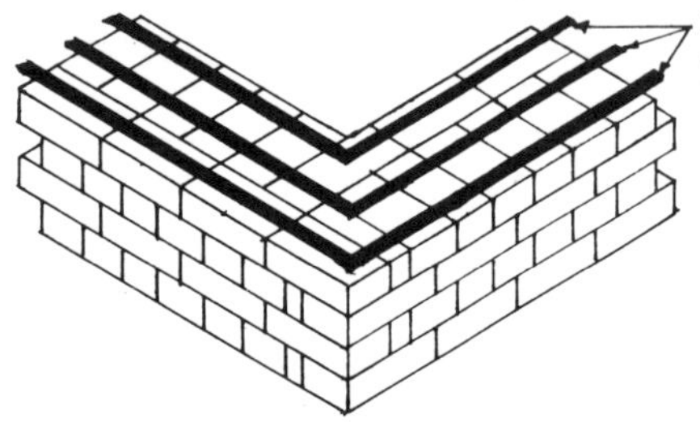

Figure 3.8 Horizontal reinforcement with hoop-iron

Rod Reinforcement

This is used vertically to strengthen walls of reduced thickness and to resist lateral stresses. Rods are used with diameters of 12–20 mm, depending on the situation and the requirements of the structural engineer. It is common practice to use this method of reinforcement within concrete pockets formed in the thickness of the wall (figures 3.10 and 3.11) or in concrete walls with brick facings. In these positions wires and stirrups are often used to increase the longitudinal stability of the reinforcement.

In modern construction the use of perforated bricks, with perforations designed to accommodate the vertical rod reinforcement (figure 3.9), provides a wall that is reduced in both thickness and weight but with sufficient strength for the situation in which it is required.

Illustrations of reinforcing with rat-trap bond, wall ties, diagonal bond and brick lintels are given in figures 3.12–3.15.

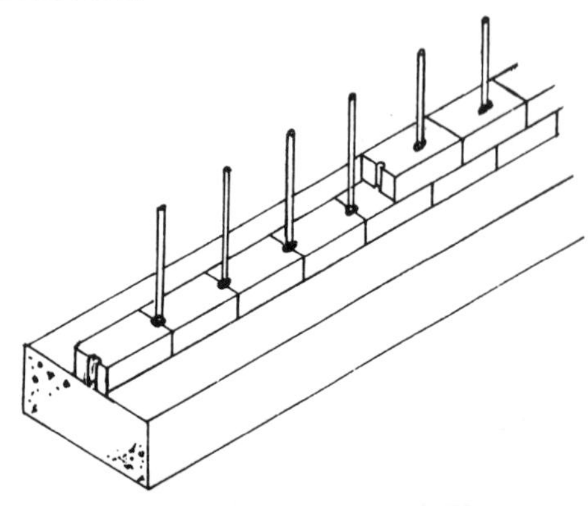

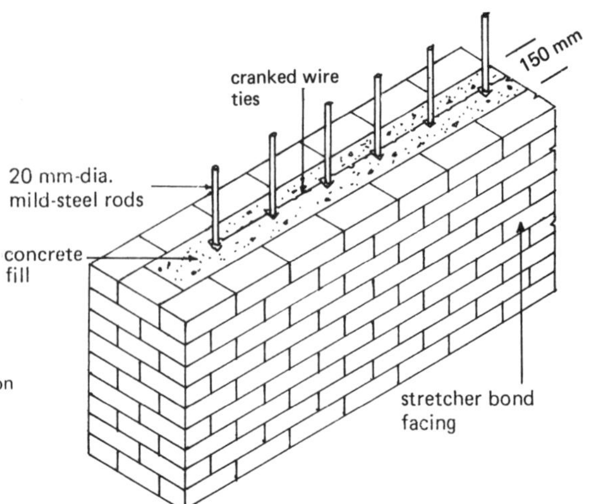

Figure 3.10 Thick walling with vertical reinforcement

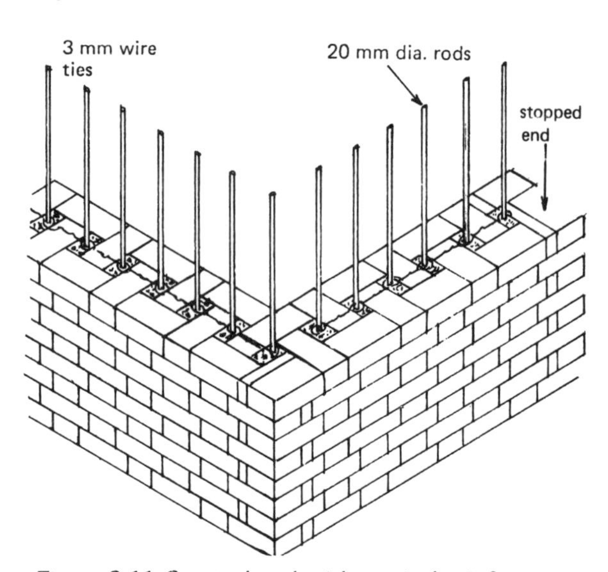

Figure 3.11 Quetta bond with vertical reinforcement

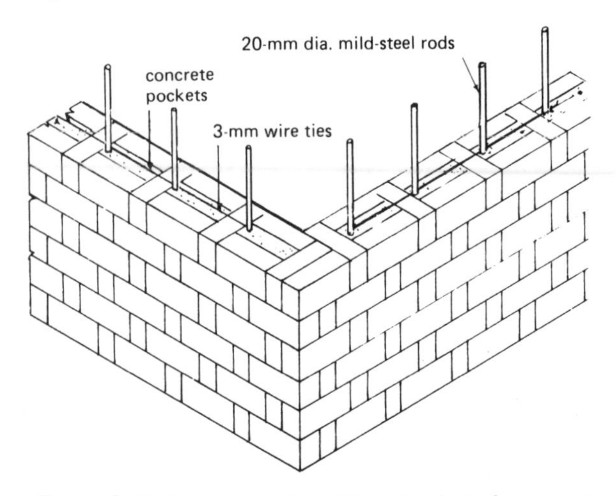

Figure 3.12 Rat-trap bond with vertical reinforcement

POSITIONING OF REINFORCEMENT

Recommended positions for the reinforcement, when walls are to be reinforced above openings, are given in table 3.1. (See also figure 3.17.)

Table 3.1 Recommended Positions for Reinforcement above Openings

Clear span of opening (m)	Number of courses reinforced	Minimum height of brickwork above opening (mm)
1.2	2	600
1.3	3 (figure 3.17)	750
1.8	3	900
2.0	4	1075

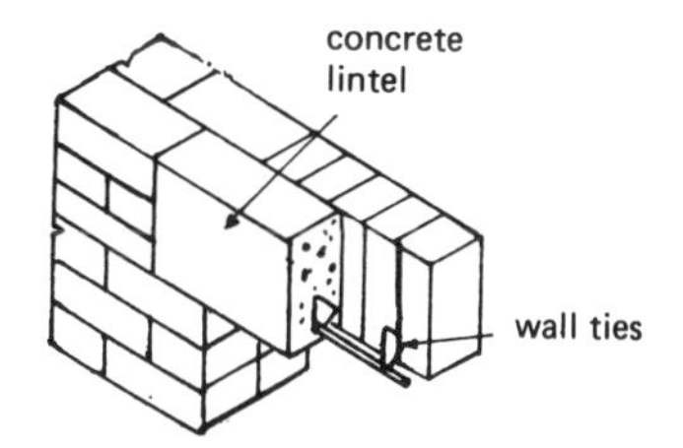

Figure 3.13 Wall ties fixed on 20 mm rod on every alternate course

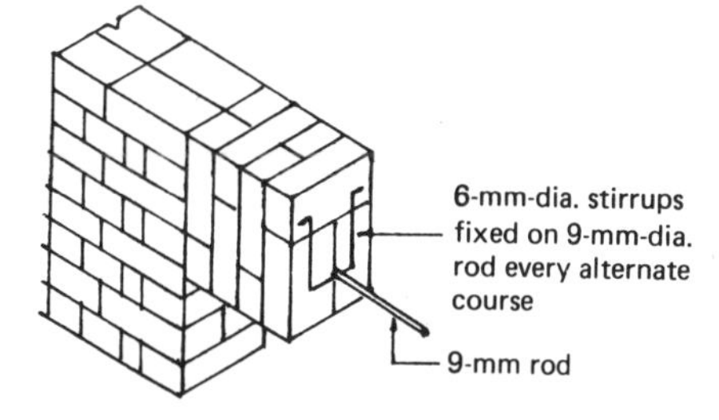

Figure 3.14 A reinforced brick lintel

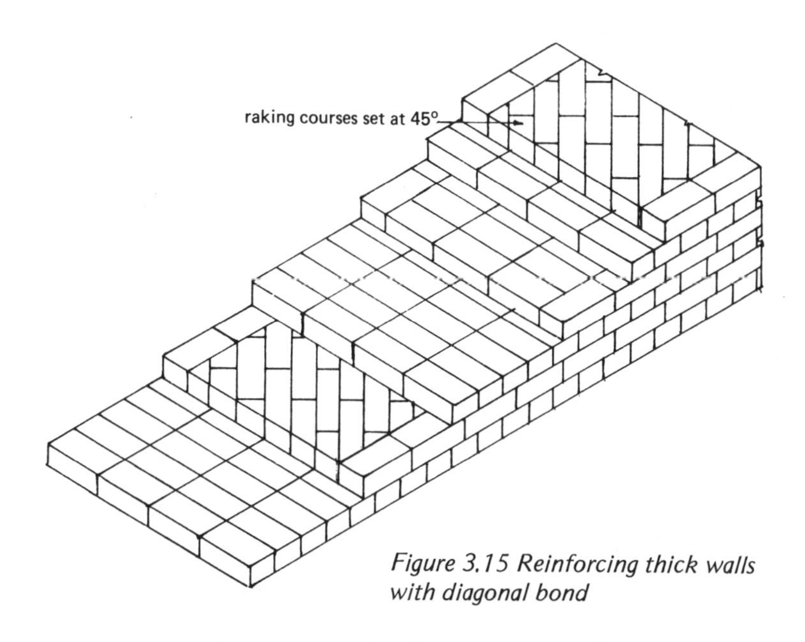

Figure 3.15 Reinforcing thick walls with diagonal bond

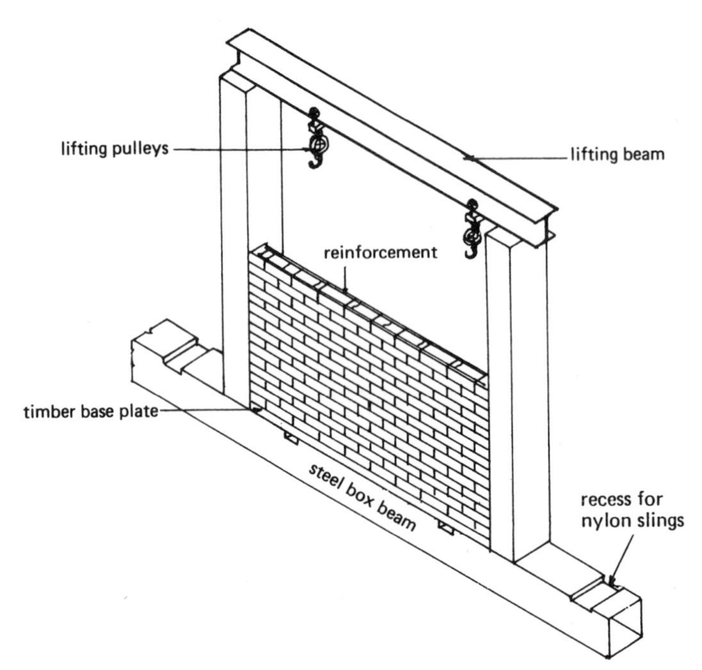

Figure 3.16 Jig assembly for building prefabricated brick panels

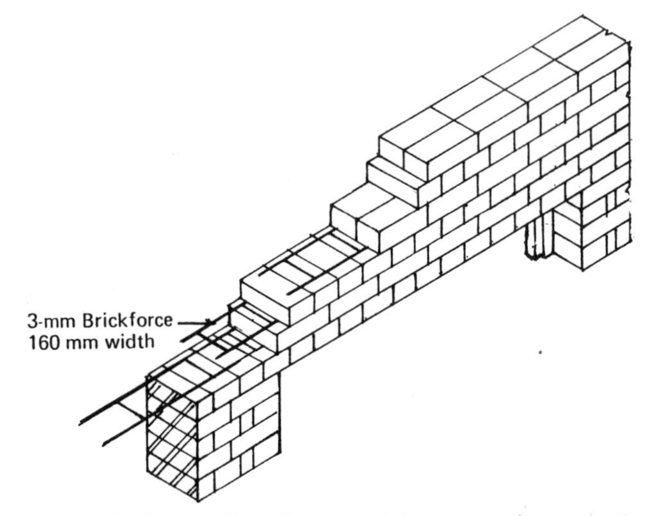

Figure 3.17 Horizontal reinforcement in courses above openings

4
STEP CONSTRUCTION

Steps are formed to provide access from one level to levels above and should be designed to prevent unnecessary fatigue or physical discomfort. Steps may be formed in brickwork, concrete or a combination of both materials.

To understand the construction of any type of step work it is necessary to have a knowledge of the following definitions.

Riser The vertical face between consecutive treads.
Tread The upper surface of a step.
Rise The vertical distance between consecutive steps or between step and landing.
Pitch line The notional line that connects the nosings of all treads.
Pitch The angle formed between pitch line and the horizontal.
Nosing The front edge of a step or tread.
Going The horizontal distance on the plan between the nosing of a tread and the nosing on the tread above.
Flight Part of a stairway, or ramp, which may consist of a step or consecutive steps.
Parallel Steps of uniform width.
Width The distance between the nosing and the face of the riser (figure 4.1).

REQUIREMENTS FOR BRICK AND CONCRETE STEPS

All steps must be built or formed to comply with the Building Regulations 1985 and the minimum going, pitch, maximum and minimum rise and width are now determined by the type and use of the building. The dimensions given in figure 4.1 refer to domestic buildings only.

Steps can be supported by walls or concrete (figures 4.2–4.4). They can be built in during construction or fixed at a later date.

Concrete Steps

When concrete steps are built into the walling, either during construction or later, it is necessary to provide a gauge staff, which should be formed to suit the brickwork courses and the height of the risers (figure 4.2). When the steps are fixed at a later date, the wall should contain recesses, formed in the brickwork or sand courses, to accommodate the concrete steps (figure 4.2).

Concrete steps, when formed as precast units, can have stepped or sloping soffits, the latter having squared seatings at each end to rest on the walling (figures 4.3, 4.5 and 4.6).

Brick Steps

Brick steps should be built of durable bricks, which should be able to resist abrasion and weather and should be set in cement mortar suitable for the type of brick used.

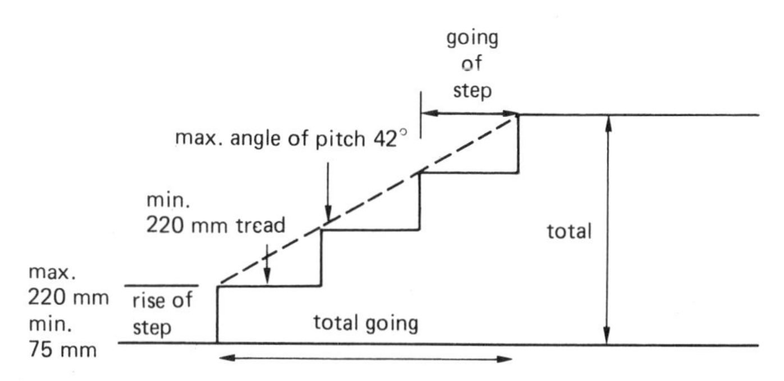

Figure 4.1 Building Regulations requirements for step construction

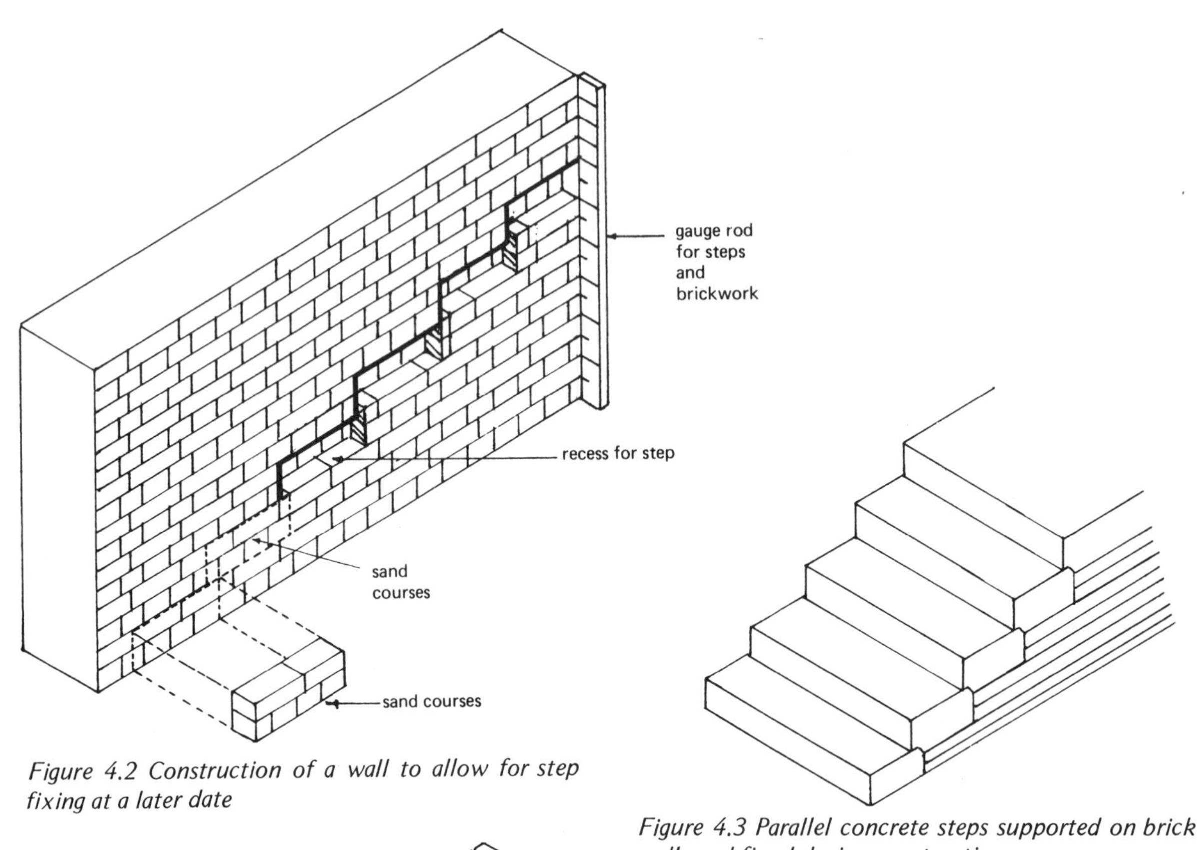

Figure 4.2 Construction of a wall to allow for step fixing at a later date

Figure 4.3 Parallel concrete steps supported on brick walls and fixed during construction

Figure 4.4 Brick steps formed between flank walls

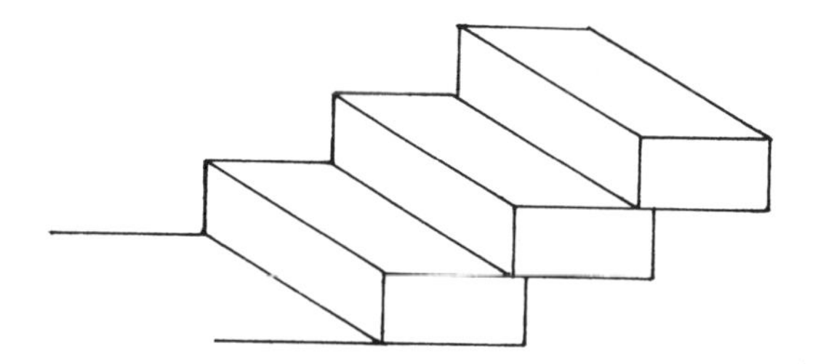

Figure 4.5 Parallel steps with stepped soffit

Balustrades

Balustrades are formed as walls or railings with minimum and maximum heights of 840 mm and 1 m respectively. They are only necessary where the total rise is greater than 600 mm, the height being measured vertically above the pitch line.

If metal balusters are used, holes will have to be prepared for these close to the ends of the treads. They are usually fastened in position with molten lead or a rapid-hardening cementitious compound.

Pyramidal brick steps are illustrated in figure 4.8.

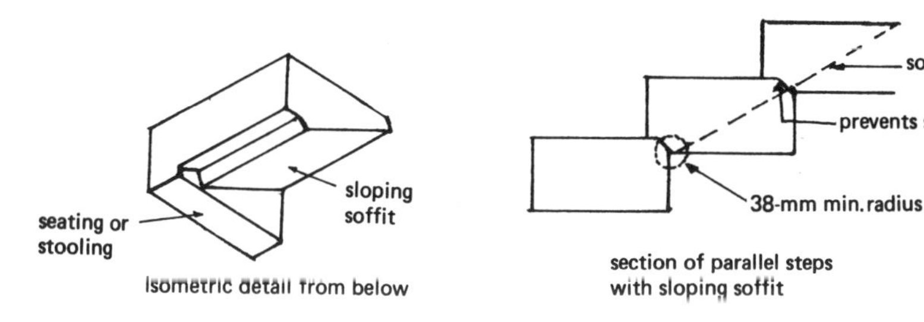

Figure 4.6

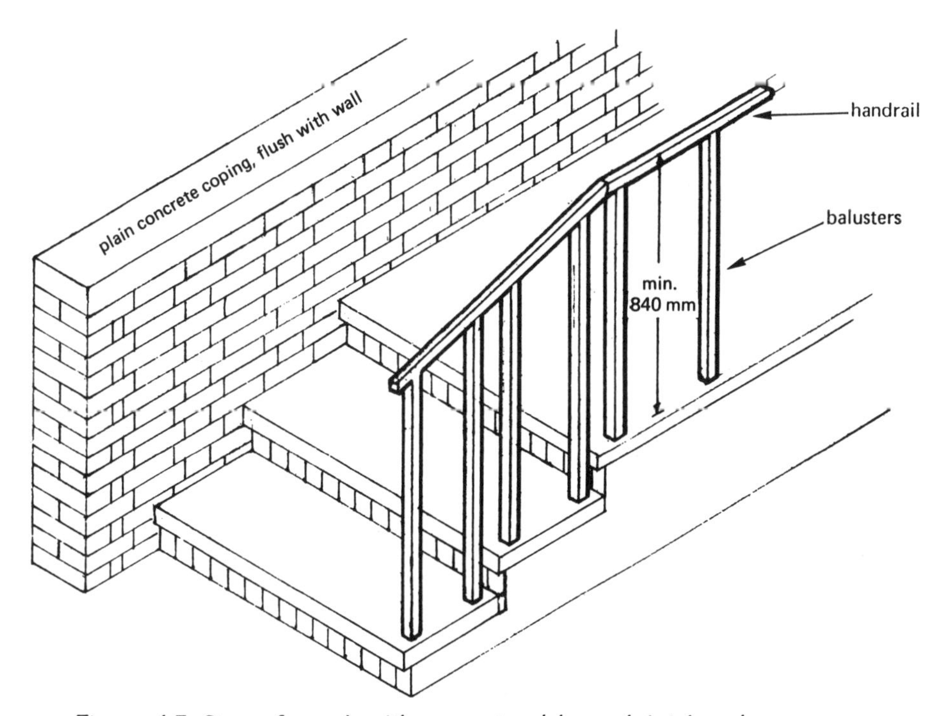

Figure 4.7 Steps formed with concrete slabs and brickwork

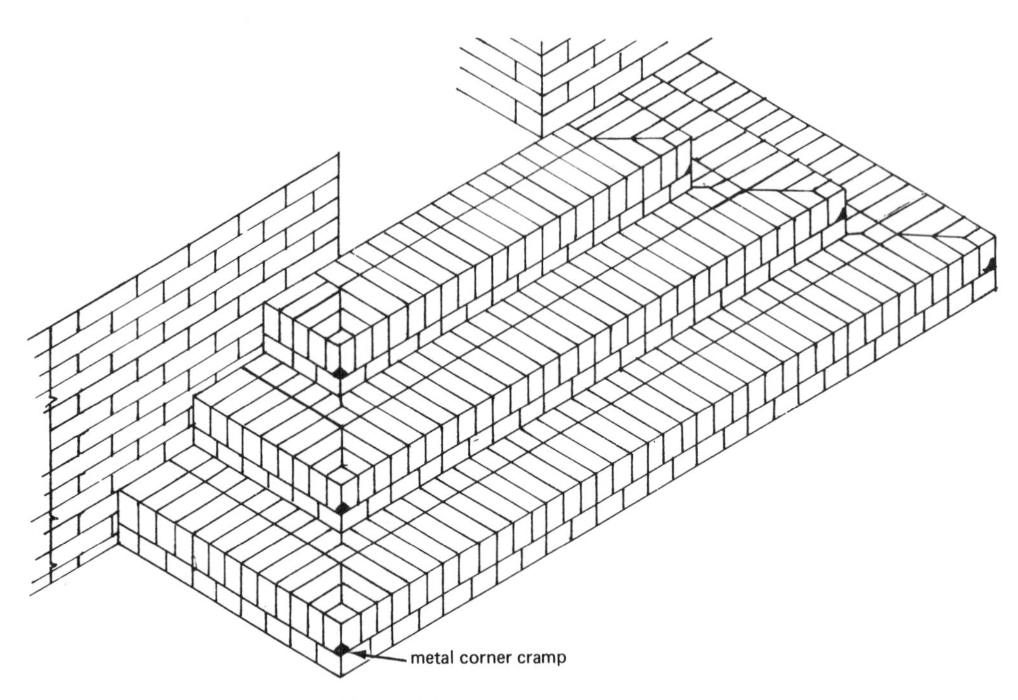

Figure 4.8 Pyramidal brick steps

5
DRAINAGE

The object of a drainage system is to convey foul, waste or surface water to the sewer or other place of disposal without danger to health. This means that the pipework must be airtight and watertight in order that both solid matter and liquid matter are removed from a building without foul odours escaping, except where this is part of the design (see figure 5.35).

Definitions of water types are as follows (figure 5.1)

Surface water The run-off of natural water from the ground surface, including paved areas, roofs and unpaved land.
Ground water In permeable ground the surface water will percolate downwards towards the water table, being held temporarily in suspension.
Subsoil water Water occurring naturally below the ground surface, the depth varying with the season.
Waste water The discharge from lavatory basins, baths, sinks, etc., that is, water not classed as surface water and not contaminated with soil water.
Soil water The discharge from soil appliances such as water closets, urinals, etc.

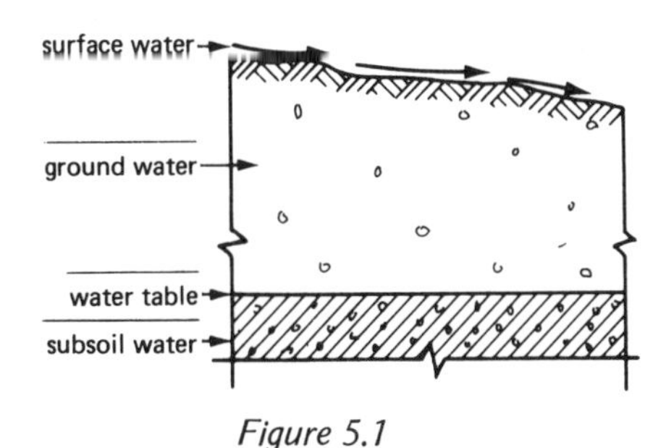

Figure 5.1

SUBSOIL DRAINAGE

The Building Regulations 1985 state that wherever the dampness or position of the site of a building renders it necessary, the subsoil must be effectively drained as required to protect the building against damage from moisture.

Furthermore whenever excavation work causes a subsoil drain to be severed, adequate steps must be taken to secure the continued passage of subsoil water through this drain or otherwise to ensure that no subsoil water entering such a drain causes dampness of the site of the building.

Drainage of subsoil water may be necessary for any of the following reasons

(1) to prevent surface flooding and thus improve conditions for building
(2) to lessen the amount of dampness occurring in foundation brickwork
(3) to prevent foundation trenches from becoming waterlogged
(4) to increase the stability of the subsoil and the ground surface
(5) for agricultural purposes
(6) to lessen the humidity that can occur when buildings are erected on damp sites.

Systems

The following are the usual methods of carrying out subsoil drainage, depending on the location of the site and the conditions prevailing.

Natural

Trenches are excavated and pipes laid to follow the natural contours on the site with branch drains discharging into the main drain as necessary (figure 5.2).

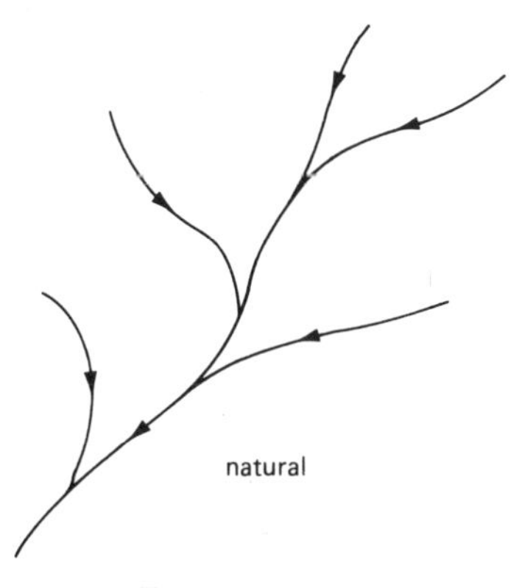

Figure 5.2

Herringbone

On a relatively large site the main can be laid down the centre, receiving the discharge from branches that are not longer than 30 m (figure 5.3).

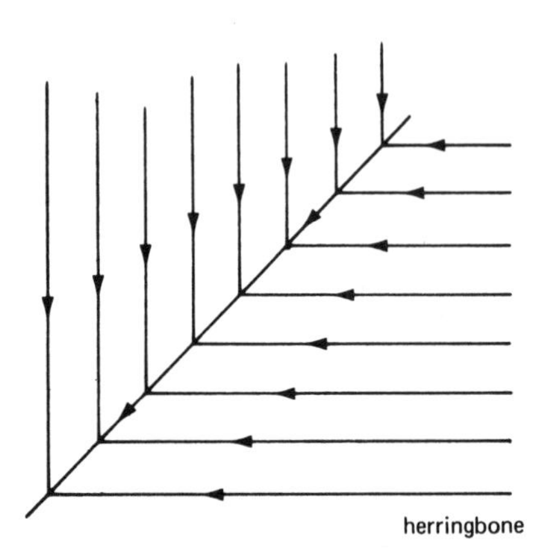

Figure 5.3

Parallel

On a smaller site, branches from one side only may be sufficient (figure 5.4).

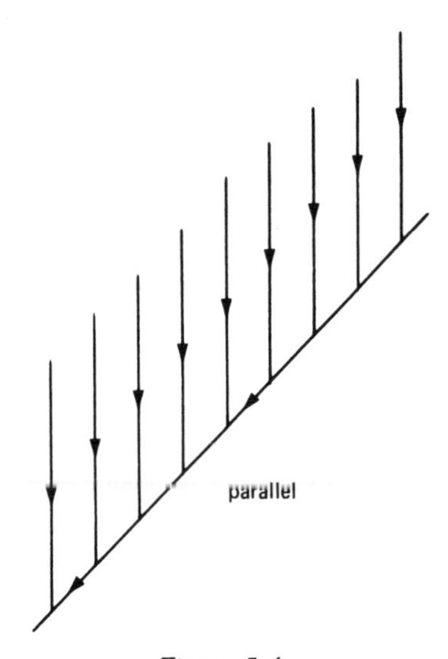

Figure 5.4

Fan

A series of small branches converges on a larger drain situated at or near the site boundary (figure 5.5).

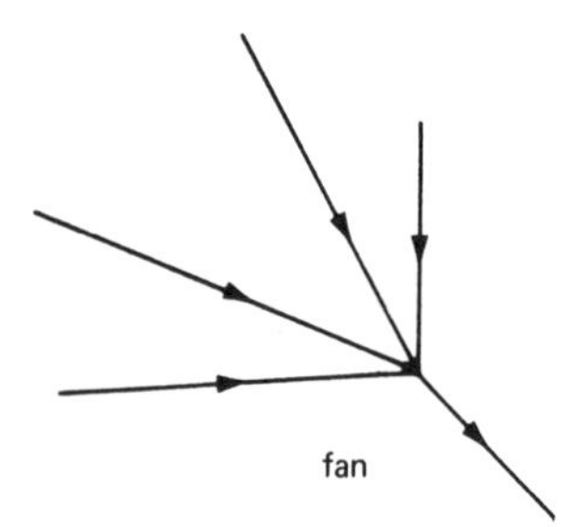

Figure 5.5

Moat or Cut-off

Subsoil drains are laid around a proposed building, thus intercepting the flow of subsoil water (figure 5.6).

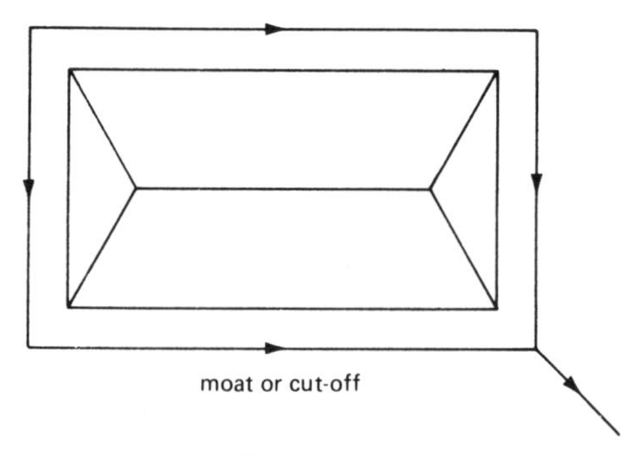

Figure 5.6

Main, with Collectors and Laterals

On a large site the subsoil drainage may consist of a series of collectors and laterals discharging to a main subsoil drain, probably of 150 mm diameter, or to a moat (figure 5.7).

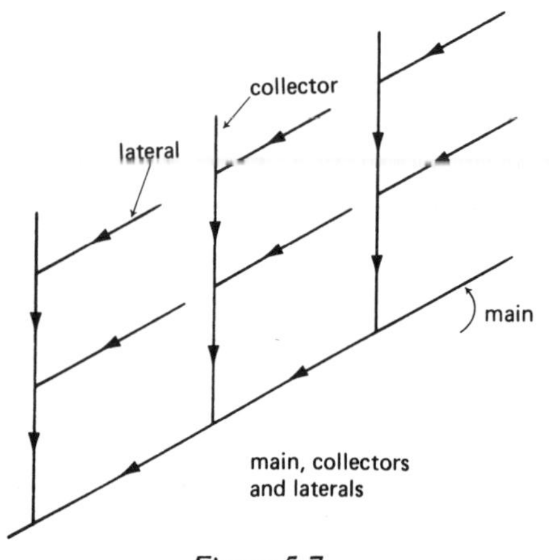

Figure 5.7

Pipes

Clayware

Clayware pipes are circular in section, butt-jointed and vary from 64 mm to 300 mm in diameter. Lengths are from 300 mm to 600 mm (figure 5.8*a*).

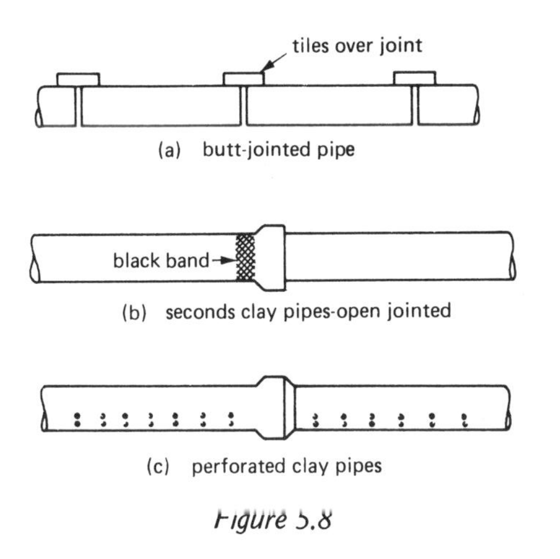

Figure 5.8

Clay Drainpipes

Types of clay drainpipe available include butt-jointed, 'seconds' and half and fully perforated pipes (figure 5.8*c*). Butt-jointed pipes normally have a tile over the joint to prevent direct ingress of silt, and seconds (identifiable by the black band) are laid open-jointed with level inverts (figure 5.8*b*). Where the half-perforated pipes are used the perforations are usually kept in the underside so that the pipes will not silt up due to fine particles moving downwards with ground water.

Other kinds of pipe for subsoil drainage include concrete porous, perforated vitrified, pitch-fibre perforated and polythene slotted pipes. They are usually laid to follow the fall of the land where possible and in trenches to a minimum depth of 600 mm. After surrounding the pipes with hardcore or rubble the pipework is then covered with an inverted layer of turf, which acts as a filter, and the trench is filled with permeable soil (figure 5.9).

Alternatively, the trench can be filled to just below ground level with hardcore (figure 5.10), or a french drain can be used, which usually consists completely of hardcore (figure 5.11).

The outfall from a subsoil drainage system should be led to a stream if possible, to soakaways where the ground is sufficiently permeable, or to a waste-water drain via a catchpit (figure 5.12).

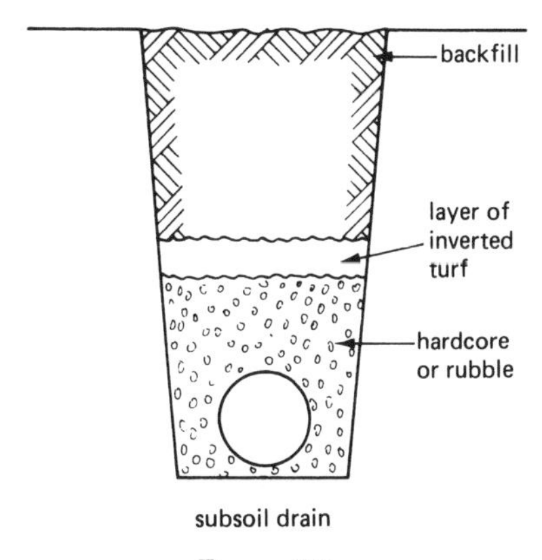

Figure 5.9

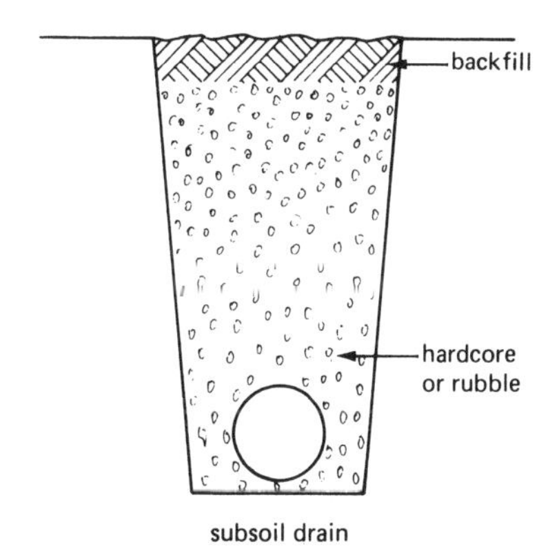

Figure 5.10

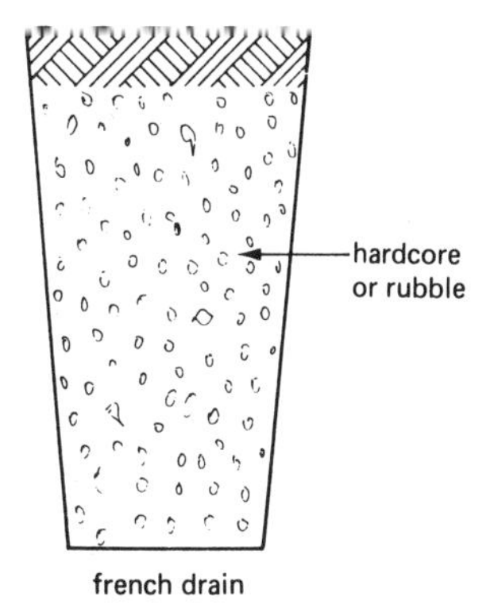

Figure 5.11

Soakaways

Soakaways are pits that are excavated in permeable soil and suitably prepared to receive ground water.

(1) The pipework is led into the pit, which is filled

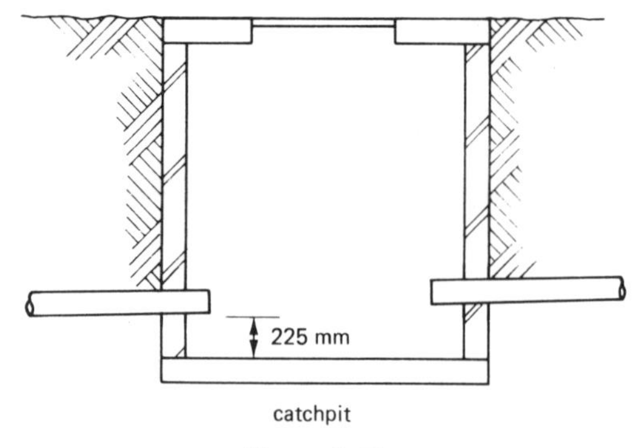

Figure 5.12

with large hardcore and covered with a layer of weak concrete. This method is cheap but provides only limited storage space (figure 5.13).

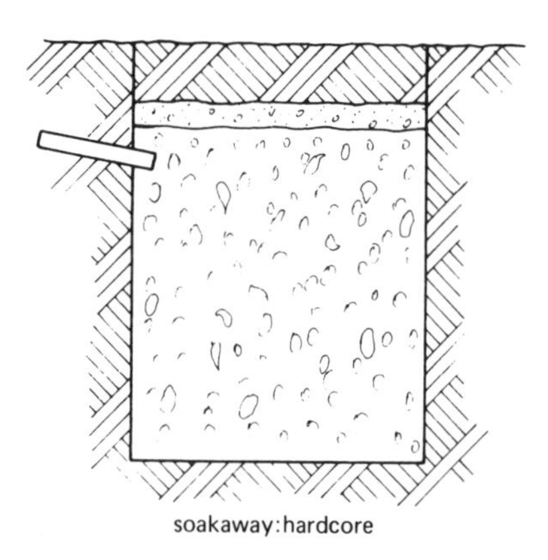

Figure 5.13

(2) Precast concrete sections are placed where required and excavation is carried out from the inside, the sections sinking under their own weight. These have a large, easily calculated capacity but costs are incresed (figure 5.14).

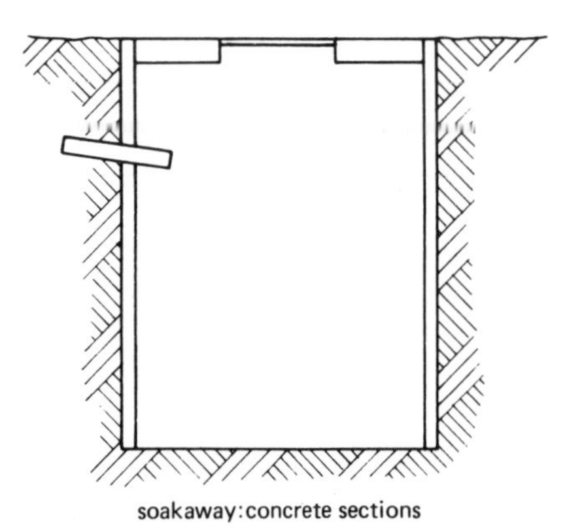

Figure 5.14

(3) An older method is to line the excavated pit with dry brickwork or stone walling, which cuts down on costs, has an easily calculated capacity but is not such a permanent job (figure 5.15).

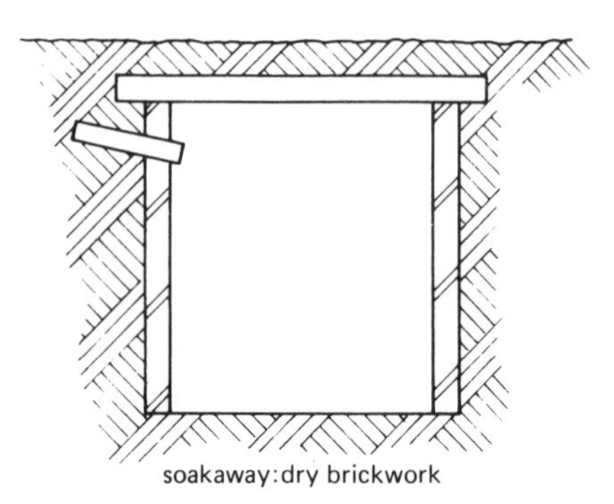

Figure 5.15

Whichever method is decided on it is vital to ascertain that the subsoil is permeable, otherwise the soakaway will become a well. When the nature of the subsoil is not known, trial pits should be excavated on site and the rate of percolation noted. It is often the case that the nature of the subsoil alters with depth and that permeable ground exists below clay, etc. In figure 5.16, for example, it would be virtually useless to dig a soakaway less than 1.5 m deep.

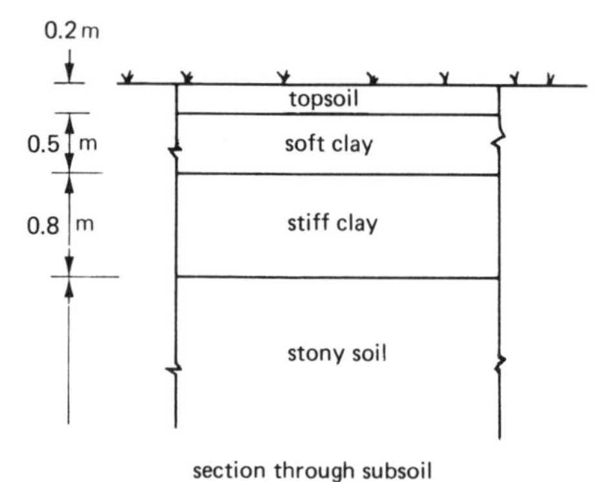

Figure 5.16

Connection to Waste-water Drains

Any silt being carried along the pipework must not be allowed into a waste-water drain and, therefore, a catchpit is installed before the connection is made. This is a brick chamber similar to an inspection chamber but containing neither channel pipes nor benching (figure 5.12). The outlet is kept about 225 mm above the top of the concrete base and the catchpit must be cleaned out at intervals as required.

Rainwater, Waste and Foul Water

All domestic buildings must be provided with efficient drainage systems in order to dispose of rain, waste and foul water. The system must discharge into a main sewer, a septic tank or a cesspool, depending on the availability of these alternatives. The sewer is laid and maintained by the local authority, usually below the road or footpath, and the house system must be connected to this either at an inspection chamber (see page 64) or between inspection chambers by means of a saddle (see page 56) fitted to the cheek of the sewer.

Septic tanks and cesspools are outside the scope of the craft syllabus but will be covered in an advanced volume.

GENERAL PRINCIPLES OF DRAINAGE

(1) The drainage layout should be as simple and direct as possible.
(2) Materials used should be hard, smooth, non-corrosive and true in shape.
(3) Pipes should be laid to falls to give a self-cleansing velocity. This is generally considered to be a flow of between 0.75 and 3 m/s. While Maguire's rule is to some extent outdated it gives a flow, depending on the type and condition of the pipe, of 1.375 m/s, that is, where a 100 mm pipe is laid at a fall of 1 in 40, a 150 mm pipe is laid at a fall of 1 in 60, and a 225 mm pipe is laid at a fall of 1 in 90.
(4) All joints must be airtight, watertight and free from internal obstruction.
(5) Lines of pipes between inspection chambers are to be as straight as possible, both horizontally and longitudinally.
(6) All inlets to drains must have a water seal of at least 50 mm, except soil and ventilation pipes (see page 63).
(7) Branches should be kept as short as possible.
(8) The greatest volume in flush should be at or near the topmost point of the drain where possible.
(9) All junctions should be made with the flow.
(10) Adequate means of access and inspection must be provided. Inspection chambers must be placed
(a) at each point where there is such a change of direction or gradient as would prevent any part of the drain being readily cleansed without such a chamber
(b) on a drain or private sewer within 12.5 m from a junction between that drain or private sewer and another, a private or public sewer, unless there is an inspection chamber situated at that junction
(c) at the highest point of a private sewer unless there is a rodding eye at that point
Note The maximum distance between manholes on a drain or private sewer is 90 m.
(11) Pipes must be laid at depths to prevent accidental disturbance or be adequately protected by haunching or be surrounded in concrete.
(12) Waste pipes to ground floors must discharge below grating level but above the water seal.
(13) Vent pipes are to be of sufficient height and to be fitted at the top with a durable cage to prevent ingress of birds, leaves, etc.
(14) Only one connection per dwelling is made to a main sewer (figures 5.28-5.33).
(15) Drains should not pass under buildings unless this is unavoidable, for example, where the length of a drain would be substantially increased or where sufficient fall cannot otherwise be obtained (figure 5.17).

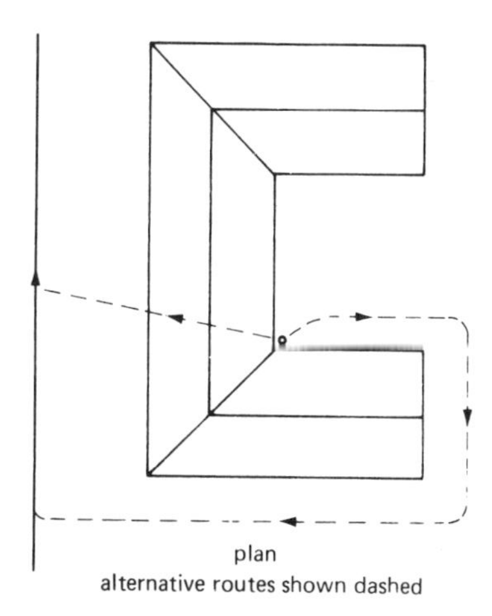

Figure 5.17

Where drains do pass under buildings precautions must be taken as necessary to prevent damage to, or loss of water tightness in, the drain or private sewer by differential movement. The principles to be followed here are

(1) Surround the pipe in at least 100 mm of granular material or other flexible filling.
(2) An inspection chamber should be placed outside at least one end of the run under the building.

(3) Where a pipe passes through a wall, either
(a) provide a 50 mm space all round the pipe, with an arch or lintel over
(b) bed a short length of pipe in the wall, forming the joints within 150 mm of either wallface. Adjacent rocker pipes should be a maximum of 600 mm long and with flexible joints (figure 5.18).

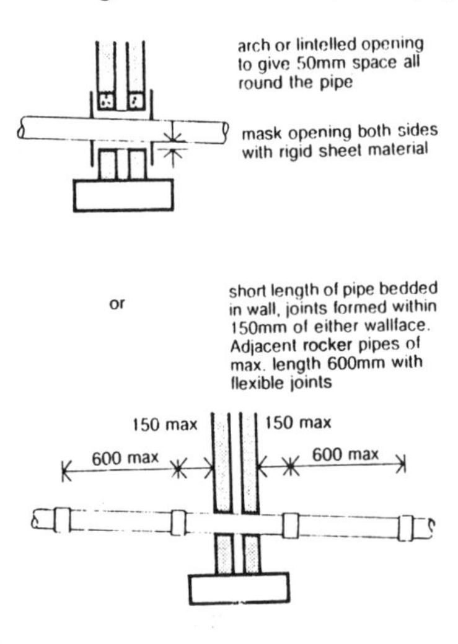

Figure 5.18 Pipes penetrating walls

Note Where, for example, an extension to a dwelling is built over an existing drain the local authority usually insists on the drain being exposed and surrounded in concrete that is continued up to the level of the natural foundation. This may not be necessary in the case of a deep drain which is usually left undisturbed. For drains that are laid close to foundations the regulations are shown in figure 5.19.

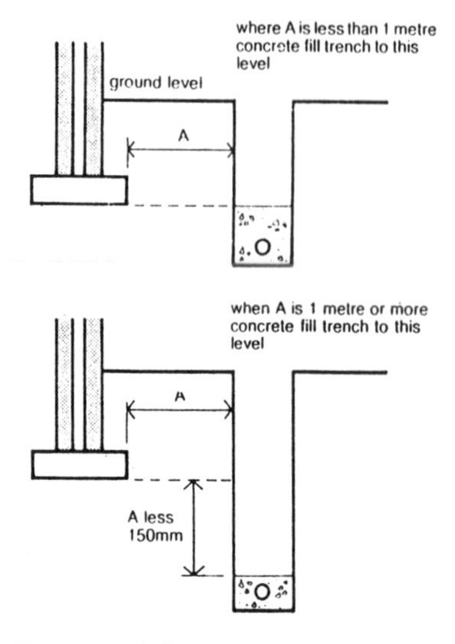

Figure 5.19 Pipes runs near buildings

Drainpipes

Pipes for drainage systems are manufactured from many materials and, while the use of cast iron has been spoken of, the craft certificate syllabus mentions those of vitrified clay, pitch fibre and uPVC.

Vitrified Clay Pipes

British Standards 65 and 540 Part 1 specify the requirements for clay pipes and fittings, with or without sockets, that are suitable for drains and sewers under two descriptions
(1) British Standard, for foul sewage and/or surface water
(2) British Standard Surface Water, for surface water only

and pipes must be clearly marked as such. Pipes and fittings of either description may be glazed or unglazed externally, internally or both.

Part 2 of the Standard specifies the requirements for flexible joints for pipes complying with Part 1.

Pipes with Sockets These have rigid or flexible joints and, while the latter cost more to purchase, they have many advantages over the more traditional rigid-jointed pipes

(1) Pipelaying is much quicker with the simple push-fit joints; thus labour costs are reduced.
(2) The joints, once made, are immediately watertight; thus testing is not held up.
(3) Joints can be made in waterlogged trenches or freezing conditions.
(4) The pipes are self-centring; mis-alignment at the joints cannot occur.
(5) The fact that the joints are flexible allows for slight distortions of the pipeline due to ground movement without loss of watertightness.

Two types of flexible joint, the O-ring joint and the polyurethane joint, suitable for socketed pipes, are shown in figures 5.20*b* and *c*, the sleeve joint (*d*) being for butt-ended pipes only.

A rigid joint is shown in figure 5.20*a*, being made with gaskin/cement mortar. This joint is described in more detail later.

A complete range of fittings is available for both rigid and flexible-jointed pipes, some of which are shown in figures 5.21 and 5.22.

When ordering channel pipes with collars, fittings must be described as left or right-handed. For example, a left-hand fitting, when viewed against the direction of flow, branches or bends to the left; and where a double spaced junction is required, the first branch from the spigot is stated first, that is, left/right; right/left (figure 5.21).

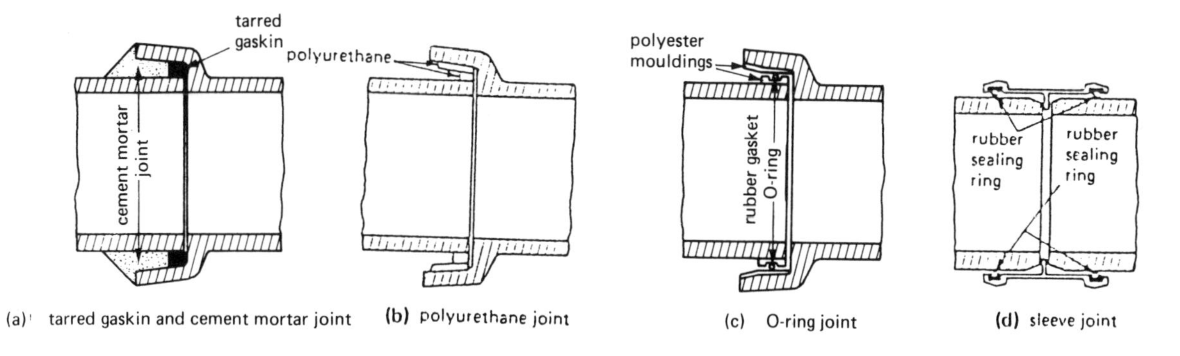

Figure 5.20 Rigid and flexible joints

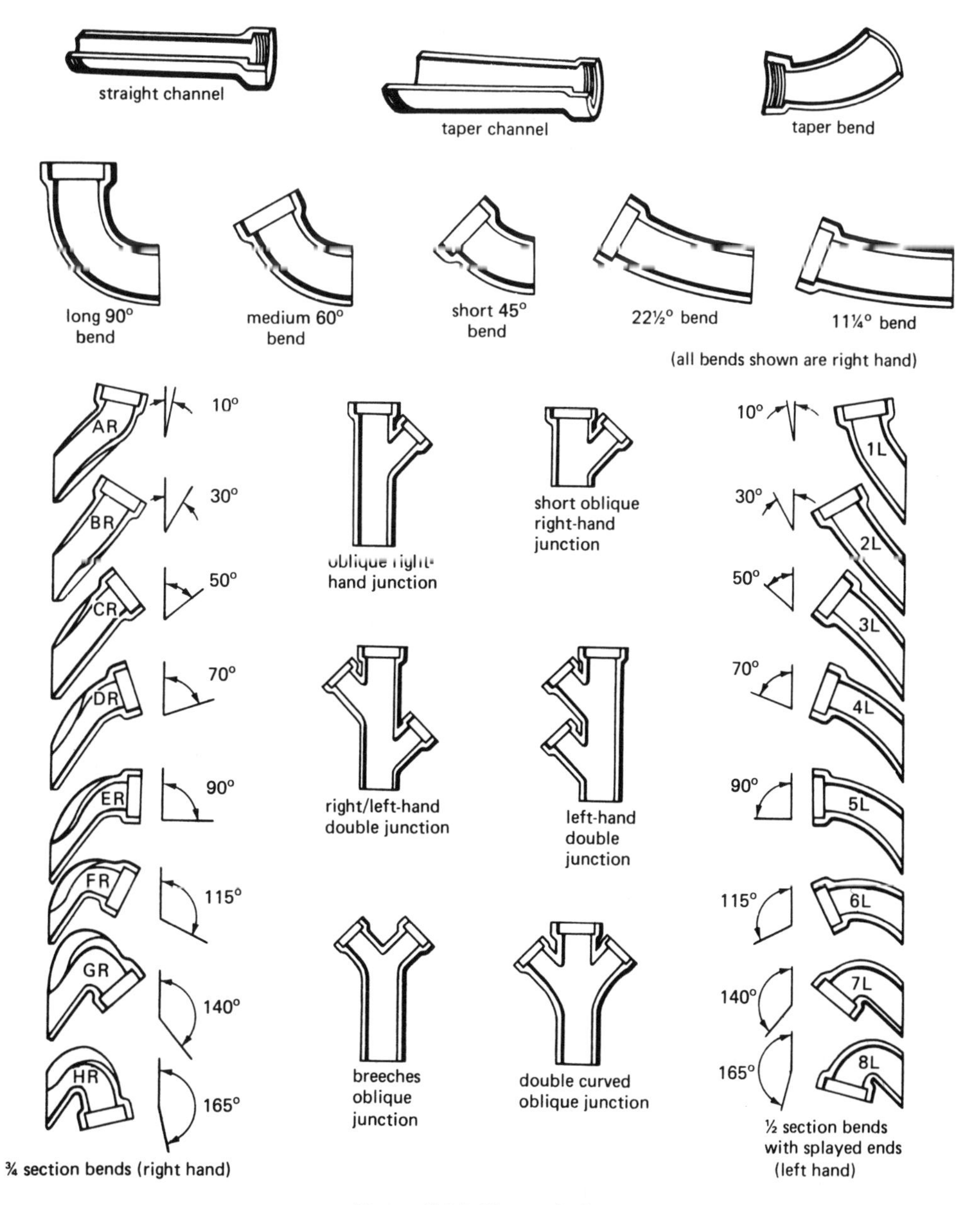

Figure 5.21 Channel pipes

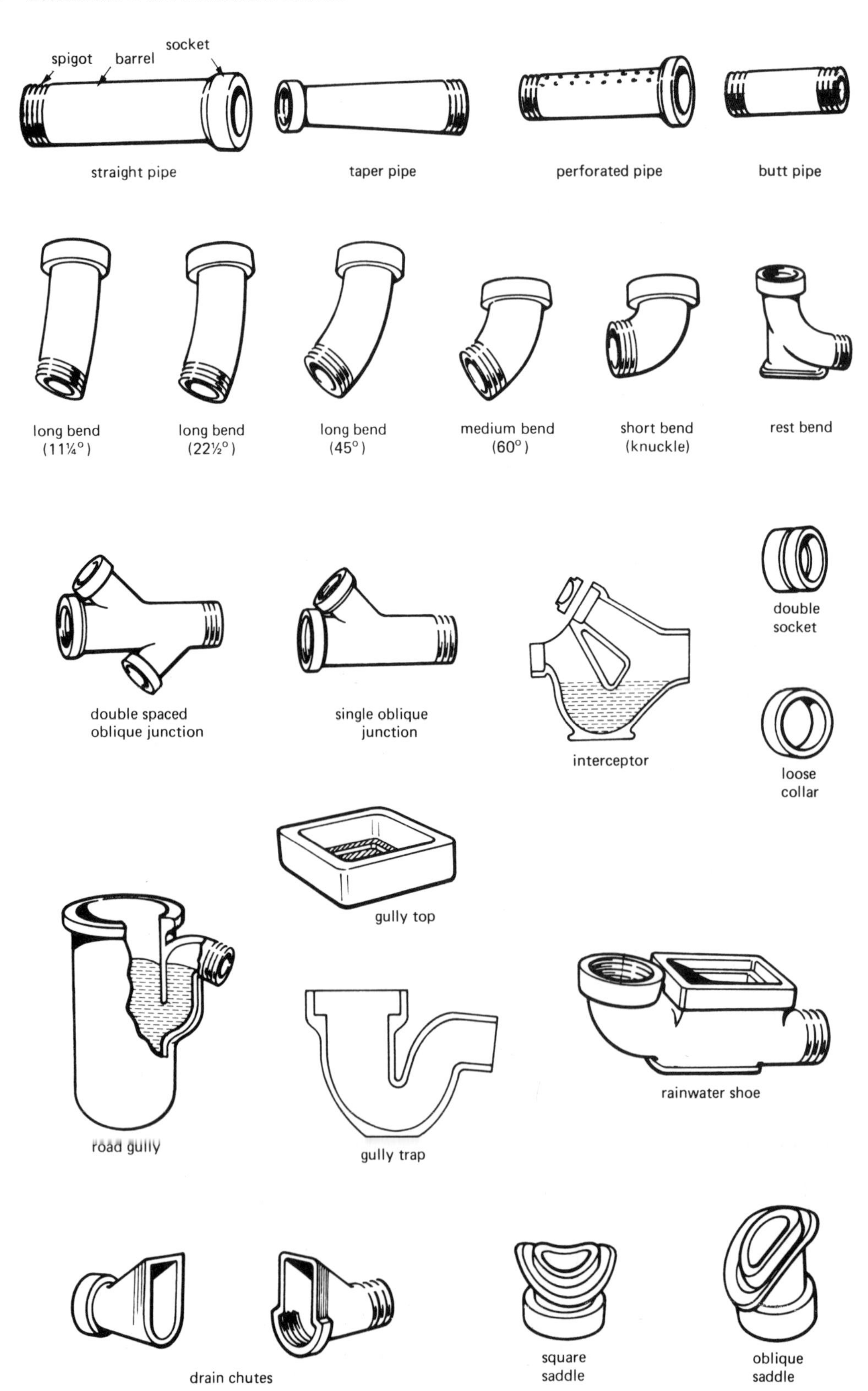

Figure 5.22 Drainpipes and fittings

Pipes without Sockets In the Hep-sleeve system, butt-ended pipes are jointed with a polypropylene coupling and rubber sealing ring (figure 5.20*d*).

After a check has been made that the pipe end and the coupling are clean, the pipe is placed vertically on a clean hard surface and the top end is lubricated. A little downward pressure on the coupling easily forces it into place against the central stop. The first pipe in the trench should be butted up to a stopboard and the lubricated end of the next pipe is forced into the collar of the first. Little waste occurs with this system since cut pipes can be utilised after trimming the cut end as necessary. As with collared pipes, a wide range of fittings is available.

uPVC Pipes (unplasticised polyvinyl chloride)
uPVC drainpipes are suitable for all foul and surface water drainage and a complete range of pipes and fittings is available. Some of the advantages claimed by uPVC pipe manufacturers include:

(1) Lightness and ease of handling. Straight pipes are available in 3 m and 6 m lengths, the latter weighing 10–14 kg (100 mm dia.).
(2) Cheaper laying costs. The material has a high impact resistance, eliminating site breakages. Long lengths mean fewer joints and any off-cuts are re-usable.
(3) No concrete bedding or haunching necessary. Granular bedding such as pea gravel is preferred, effecting further savings.
(4) Immediate testing. Rubber O-ring push-fit joints are used.
(5) Exceptionally smooth bore means an improved flow rate. Lower gradients are possible.
(6) Flexible. Little risk of failure due to ground movement.
(7) If brickbats or stones fall into trenches they will usually bounce off the pipes, rather than cause breakages.

The O-ring joints are assembled as follows:

(1) Ensure that both spigot and socket are free from grit or mud etc.
(2) Insert a clean rubber ring into the O-ring recess.
(3) Apply the special lubricant to the exposed surface of the rubber ring within the socket and to the chamfered area of the spigot end.
(4) Bring spigot and socket into alignment and push the spigot home (figure 5.22).

Pitch-fibre Pipes

This type of pipe is flexible, light in weight and generally available in lengths between 2400 and 3040 mm. While pitch-fibre pipes are considered unsuitable

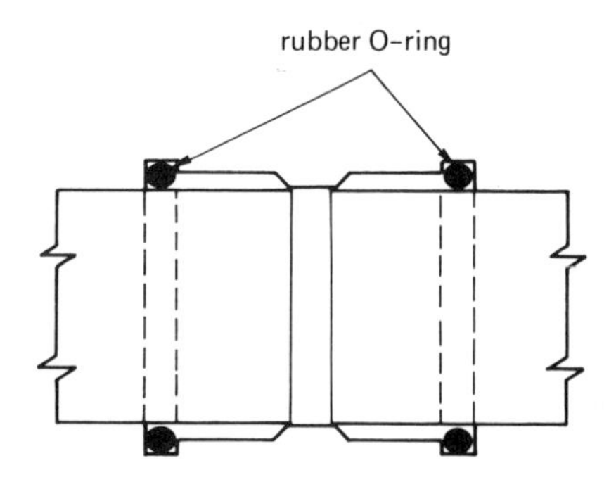

Figure 5.23 O-ring joints in uPVC drainpipes

for continuously running hot water such as the discharge from laundries, or for waste containing pitch solvents (petrol, oil and fat), they have the following advantages over rigid-jointed clay pipes

(1) their flexibility makes them suitable for ground liable to differential settlement
(2) concrete bedding, haunching and/or surround is very rarely required
(3) there is no waste, cut pipes can be used
(4) pipelaying can continue in freezing weather and waterlogged trenches
(5) immediate testing on completion is possible
(6) the pipes can be laid at lower gradients than pipes of certain other materials, for example, 100 mm pipe is laid at a gradient of 1 in 85.

Bedding

Where the subsoil is suitable, for example, where it is free draining coarse sand or loam, the practice is to replace the soil evenly, consolidate it, and bed the pipes down on this, but granular materials are mostly used for this purpose, for example, pea gravel or broken stone of maximum size 19 mm. Clay and chalk, which are affected by percolating water, should not be used for either bedding or sidefilling, and bricks, etc. must never be used to pack pipes into line.

Laying the Pipes

The pipes are laid directly on the bedding material and the joints are made with either a straight (sleeve) coupling or a snap-ring joint (figures 5.24 and 5.25).

Straight Couplings (figure 5.24)

The pipe spigots are wiped clean and a coupling is fitted hand-tight to one spigot of each. Place the first pipe in position against a suitable stopboard (figure 5.26) and fit the spigot of the second pipe into the coupling of the first. A softwood dolly (figure 5.26) is placed against the coupling of the second pipe,

which is then driven home with a 1-2 kg hammer. The amount of drive should be 6-7 mm and it is important not to overdrive. It is possible to complete a considerable length of pipe at ground level if required and to lower it into the trench afterwards.
Note When using these couplings it is necessary to support the back of a bend or a branch junction while driving (figure 5.27).

Snap-ring Couplings (figure 5.25)

The snap-ring is placed over the end of a plain-ended pipe with the flat of the ring against the pipe, and the coupling is pushed on to this, forcing the ring to roll along the pipe. The ring is compressed and jumps into the required position. Pipes are jointed in a trench by pushing a pipe with a coupling fitted into the coupling on a previously laid pipe using a spade as a lever.

Pitch fibre pipes are rarely used nowadays; they are no longer produced in the United Kingdom and are not mentioned in current Building Regulations.

Pipe Bedding and Backfilling

Figure 5.28 shows two suitable methods for rigid pipes and one for flexible pipes. Each of these will satisfy Building Regulations requirements.

Figure 5.28*a* requires a high standard of workmanship. This method must not be used unless accurate hand-trimming by shovel is possible.

Figure 5.28*b* is generally suitable in all soil con-

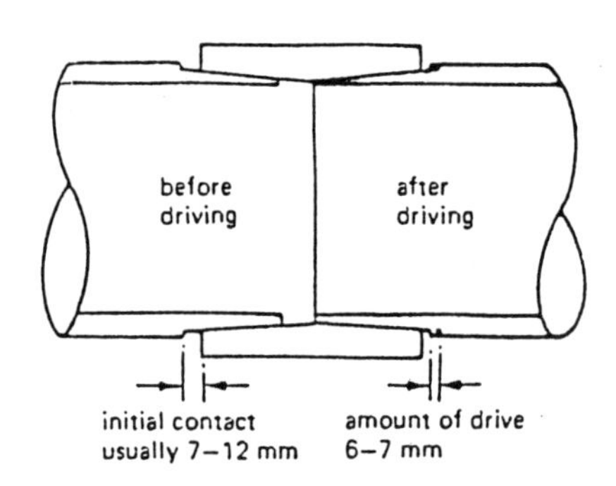

Figure 5.24 Straight coupling for tapered pipes

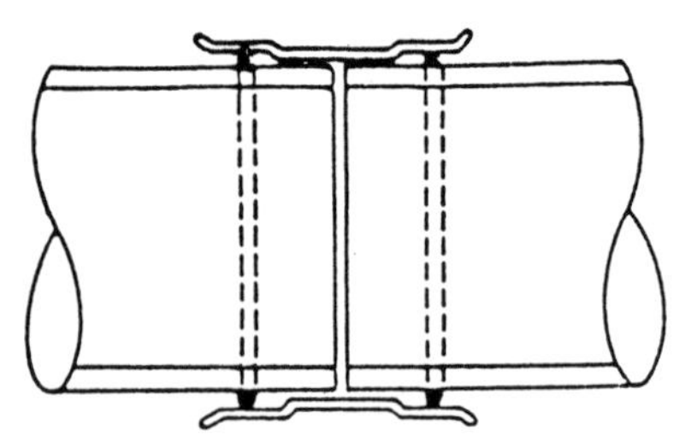

Figure 5.25 Polypropylene snap-ring joint for plain-ended pipes

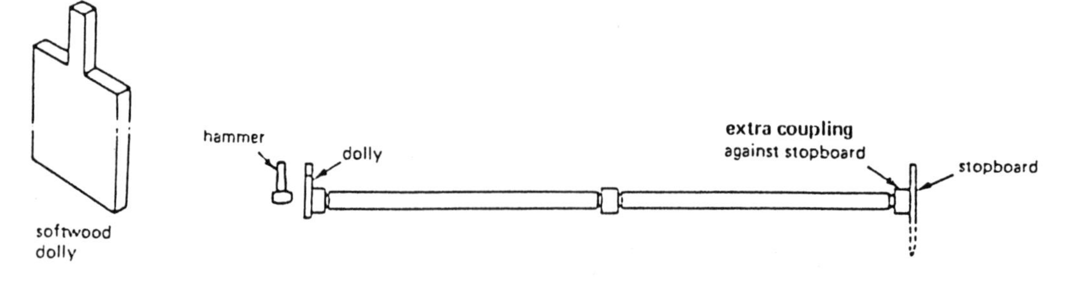

Figure 5.26

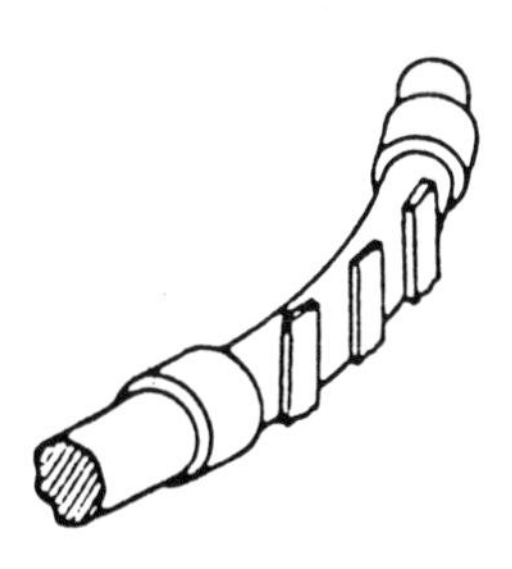

Figure 5.27 Pipes supported at bends during driving

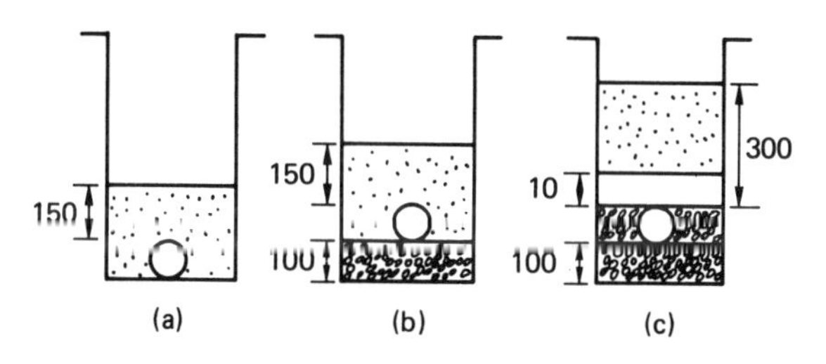

Key: Selected fill, free from stones larger than 40 mm, lumps of clay over 100 mm, frozen material, timber and vegetable matter
Granular material to BS 882
Selected fill or granular fill free from stones larger than 40 mm

Figure 5.28

ditions. It is an alternative to *a* where accurate hand-trimming is not possible.

Figure 5.28*c* is suitable for flexible pipes.

Where pipes are laid close to ground level, adequate protection must be provided. Either surround in 100–150 mm concrete or bridge over the pipes with concrete slabs on flexible filling.

DRAINAGE SYSTEMS

There are three systems of drainage, the combined, the separate and the part separate system and it is important to understand at this point that only one of these systems is used in each town or district.

Combined System (figure 5.29)

In the combined system one large system takes the discharge from top, waste and foul-water fittings. This system is found mainly in Scotland and some coastal areas where sea outfalls are used.

Advantages are

(1) the pipe layout on site is usually simple and straightforward
(2) there is no chance of connecting to the wrong sewer
(3) there is only one set of pipes and sewer to lay and maintain.

Disadvantages are

(1) the combined sewer and the treatment works must be large or they will not be adequate in wet weather
(2) storm overflows may be necessary to divert any overload to storage tanks, streams, etc., and untreated sewage may cause pollution.
(3) both foul and surface water need to be treated at the sewage works.

Separate System (figure 5.30)

With this system two completely separate sets of pipes lead to two different sewers, one conveying foul and waste water from sinks, baths, W.C.s, bidets, etc., the other taking rainwater from downspouts, paved areas, etc.

It may be possible to use soakaways where the subsoil is permeable, as shown in figure 5.30, where the soakaway at the rear is an alternative to a long run of pipes. Soakaways are placed at least 3 m from the dwelling, otherwise drain runs are laid out for the best result possible.

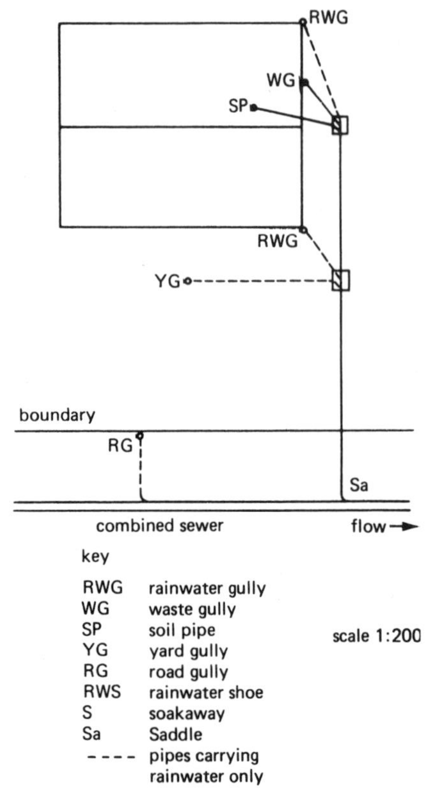

Figure 5.29 The combined system

Advantages are

(1) only foul water is treated at sewage works, thus both the treatment plant and the diameter of the sewer can be relatively small.
(2) storm overflows are not required.
(3) there is no possibility of water pollution from overflowing sewage during bad weather, since the flow is constant.

Disadvantages are

(1) there are two sets of pipes to lay and maintain
(2) there is a risk of connecting to the wrong sewer
(3) the pipe layout may be complicated with pipes crossing
(4) the foul sewer is not flushed with rainwater, therefore great care must be taken to ascertain that a self-cleaning velocity is kept to throughout.

Part-separate System (figure 5.31)

This method is a compromise between the combined system and the separate system. One sewer deals with street gullies and as much roof water as possible, the other takes foul and waste water and a small amount of rainwater, preferably, for example, via a gully at the top of the system, which will flush the drain as it flows during wet weather.

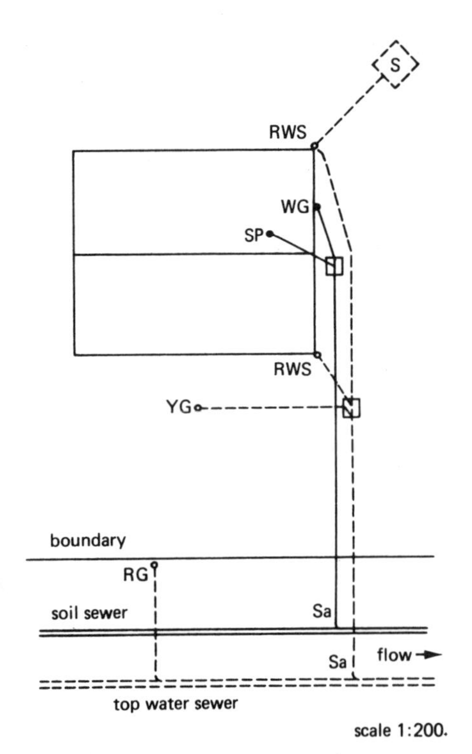

Figure 5.30 The separate system (key as for figure 5.29)

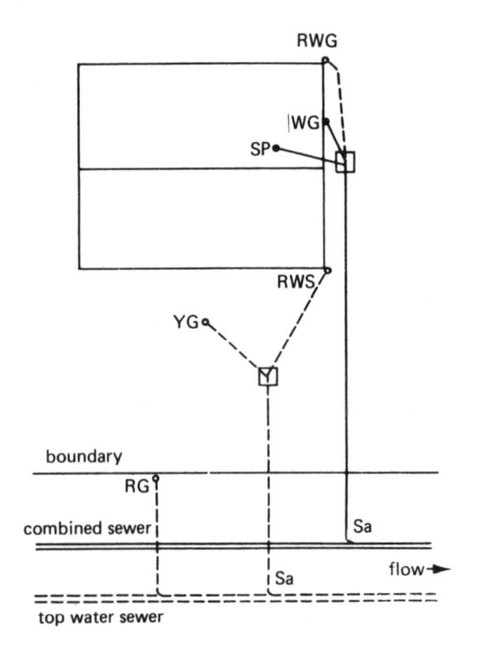

Figure 5.31 The part-separate system (key as for figure 5.29)

Advantages are

(1) the layout is usually easier and cheaper than for the separate system
(2) foul drains are flushed in rainy periods.

Disadvantages are

(1) there are two drains and two sewers to lay and maintain
(2) there are two connections to be made to the sewers
(3) the layout is costlier and more difficult than for the combined system.

Conclusions are that the combined system, despite its advantages, is considered the worst system, the separate system and the part-separate system being preferred.

Note The provision and connection of the road gully in all the systems is the responsibility of the local authority.

Measurement of Pipes

The simplest method is to measure the complete length of pipeline and specify all fittings.

Example 5.1. The Combined System (figure 5.29)

Straight Pipes

Top inspection chamber (IC) to sewer	14.6 m
Rainwater gully (RWG) to top IC	4.0
Waste gully (WG) to top IC	1.8
Soil pipe (SP) to top IC	3.2
RWG to lower IC	2.0
Yard gully (YG) to lower IC	6.0
	31.6 m

Fittings

2 rainwater gullies (to take fall pipes)

1 rest bend (foot of soil pipe)

1 waste gully (sink waste)

1 yard gully (surface water from paved areas)

1 saddle (connection to main)

1 slow bend (into top IC from RWG)

Channel Fittings

Top IC

1 left-hand double oblique junction

1 left-hand bend

Lower IC

1 left-hand double oblique junction

1 left-hand slow bend

Example 5.2. The Separate System (figure 5.30) (assuming soakaway impracticable)

Straight Pipes

Soil length

Top IC to main	14.6 m	
WG to IC	2.2	
SP to IC	2.6	
	19.4 m	(British Standard)

Surface water

Top RWS to main	21.0	
RWS to IC	2.0	
YG to IC	8.0	
	31.0 m	(British Standard Surface Water)

Alternatively, 50.4 m of British Standard straight pipes could be ordered.

Fittings

2 rainwater shoes (gullies not necessary)

1 waste gully

1 rest bend

1 yard gully

2 slow bends (for use in long length)

2 saddles (two connections)

1 slow bend (top RWS to IC)

Channel Fittings

Top IC

1 single oblique left-hand junction

1 left-hand bend

Lower IC

1 double oblique left-hand junction

2 left-hand bends

Example 5.3. The Part-separate System (figure 5.31)

Straight Pipes

Top RWG to main	19.0
WG to top IC	1.2
SP to top IC	2.6
Lower IC to main	9.0
RWS to lower IC	4.4
YG to lower IC	2.8
	39.0 m

Fittings

1 rainwater gully (connects to soil sewer)
1 rainwater shoe (connects to surface-water sewer)
1 rest bend
1 waste gully
1 yard gully
2 saddles

Cheannel Fittings

Top IC

1 double oblique left-hand junction
1 left-hand bend

Lower IC

1 breeches oblique junction

Note To give a complete specification for channel fittings required for the inspection chambers, they would have to be drawn to a larger scale, as in figure 5.37, for example.

Figures 5.32–5.34 show one further example of each of the three drainage systems.

Note It may be decided that the lower rainwater gully in the part-separate system (figure 5.31) should discharge into the top-water sewer, in which case the pipes would cross.

VENTILATION OF DRAINS

A free circulation of air must be provided through the pipes forming a domestic drainage system. This is accomplished in one of two ways, depending on the requirements of the local authority.

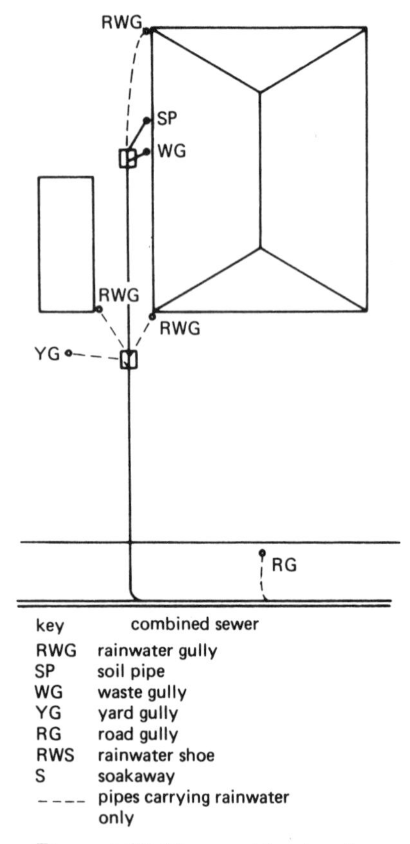

Figure 5.32 The combined system

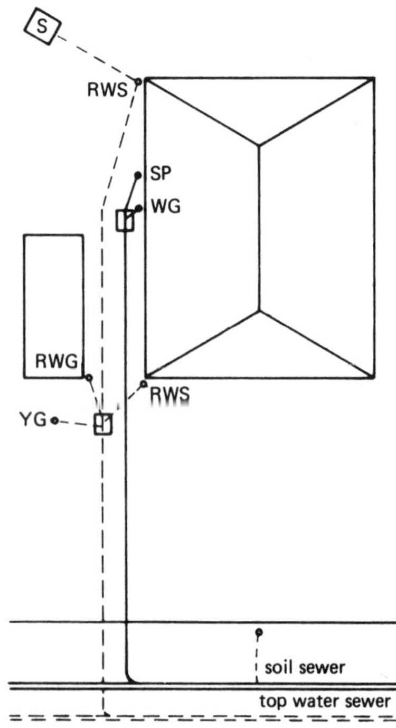

Figure 5.33 The separate system (key as for figure 5.32)

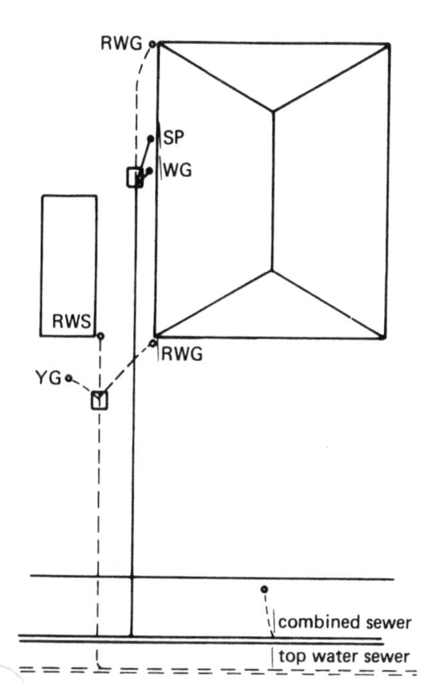

Figure 5.34 The part-separate system (key as for figure 5.32)

Ventilating without an Interceptor Trap (figure 5.35)

This is generally considered to be the best method, every drain ventilating the main sewer and thus preventing the build-up of sewer gases.

Ventilating with an Interceptor Trap (figure 5.36)

With this method, sewer gases are prevented from entering a private sewer by means of the water seal in the interceptor trap. Through ventilation of the domestic system is achieved by the provision of a fresh-air inlet at the intercepting chamber, which is installed just inside the site boundary. Fresh air is drawn into the fresh air inlet via a one-way flap, passes through the system, and is released from the top of the vent pipe, which should be provided at or near the top of the drainage system.

Disadvantages of this system include

(1) the installation of an intercepting chamber increases the cost of the system
(2) the trap itself is liable to become blocked and may require regular cleaning
(3) if the fresh-air inlet suffers damage or becomes faulty, ventilation ceases to take place
(4) some other means of preventing the build-up of gases within the sewer must be provided.

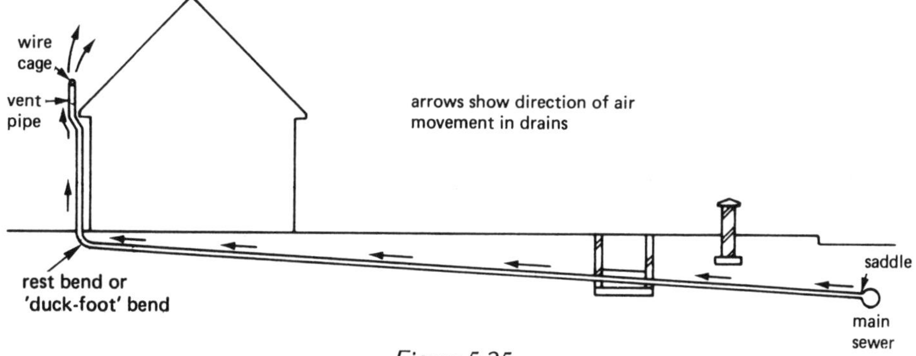

Figure 5.35

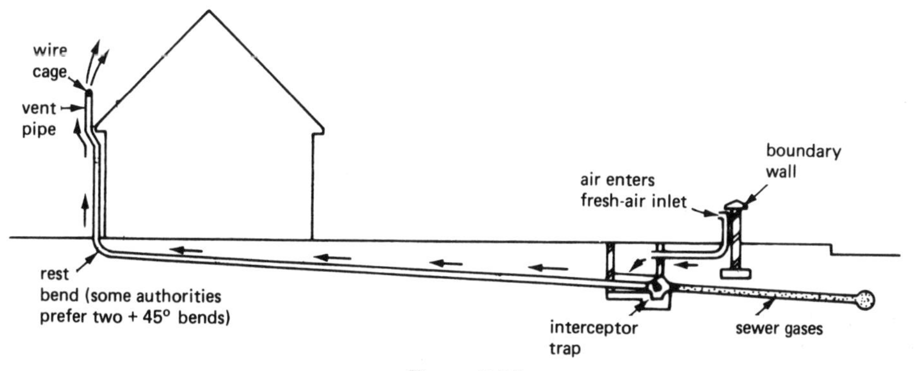

Figure 5.36

INSPECTION CHAMBERS

The purpose of an inspection chamber is to provide access for inspection and cleansing. Inspection chambers are constructed from the following materials

(1) class B engineering bricks
(2) precast concrete sections surrounded in concrete 100–150 mm thick
(3) *in-situ* concrete
(4) for surface-water drains, good quality bricks, rendered externally where deemed necessary
(5) glass-reinforced plastic.

The following notes are relevant to the construction of inspection chambers.

(1) The concrete base must be at least 100 mm thick, the thickness increasing with depth.
(2) The base can be of the same length and breadth as the overall plan area of the chamber, that is, no spread is required.
(3) The internal size varies with the depth and the number of branch drains entering: 600 x 450 mm is the minimum, increasing as the depth increases.
(4) Where bricks are used, English bond (figure 5.37*b*) is preferred to water bond (figure 5.37*c*). *Note* In water bond, the bed joints are staggered by either forming a half-course rebate around the outside of the concrete foundation, or starting the outer half-brick walling with either a course of snapped headers or a course of split bricks.
(5) Brick chambers are normally built half a brick thick where the depth to invert is less than 900 mm (figure 5.37*a*), after which one brick thick is the minimum.
(6) All pipes in inspection chambers are to be in channels discharging in the direction of the flow.
(7) A brick-on-edge arch should be formed in the brickwork over pipes more than 150 mm in

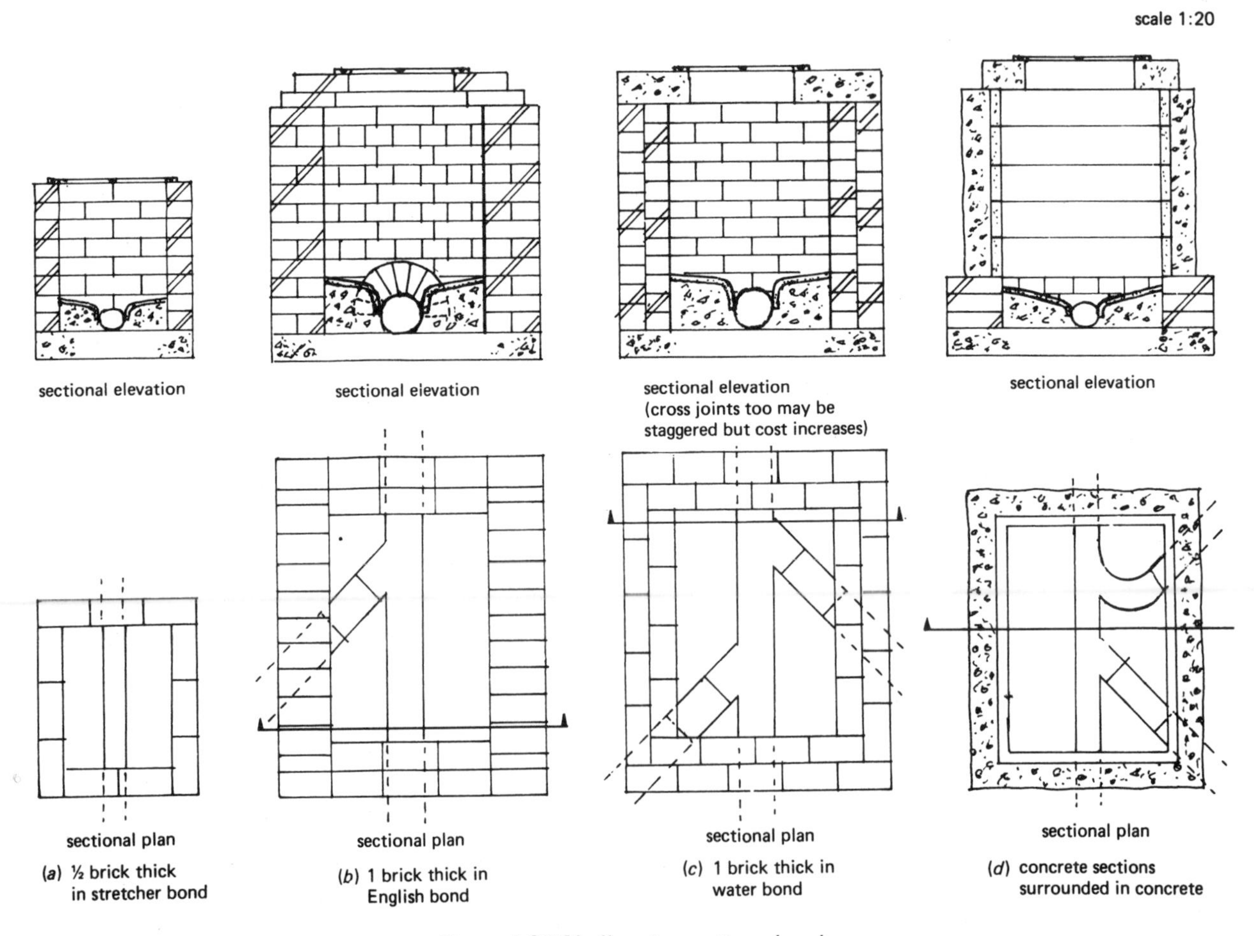

Figure 5.37 Shallow inspection chambers

diameter where the chamber is deeper than 1800 mm.

(8) Benching should rise vertically on either side of the channel to the crown of the outgoing pipe, be quickly rounded off and slope upwards towards the brickwork at a slope of about 1 in 6. The mix is to be 1:1 cement and sand, trowelled smooth.

(9) The top of the chamber must be reduced as necessary to support the cover and frame, which is usually 600 x 450 mm. This is carried out by corbelling the brickwork from one or more sides (figure 5.37*b*) or with a precast reinforced concrete slab (figure 5.37*c*). It is considered good practice to install the slab below ground level to allow for completing with two courses of bricks, so that if, at a later date, the ground level is lowered the slab need not be disturbed.

(10) Where the chamber is more than 900 mm deep, step irons (figure 5.38) should be built into the walls, the vertical spacing not exceeding 300 mm and from centre to centre 300 mm. In deep manholes a galvanised wrought-iron ladder can be used in place of step irons.

(11) At least one flexible joint on either side of an inspection chamber will help to avoid fracture in the case of ground movement.

(12) Inspection chambers are to be able to sustain imposed loads, be impervious to water and of suitable size to provide access for inspection and rodding.

(13) A removable, non-ventilating cover must be provided, with the frame normally bedded in mortar and the cover in grease to prevent the escape of obnoxious gases.

(14) Precast concrete sections are bedded on base sections having cut-outs for pipes, or three or four courses of brickwork are built to surround the pipework. The sections must be surrounded in at least 100 mm of concrete (figure 5.37**d**), depending on the depth.

SETTING OUT AND LAYING DRAINS

(1) Sight rails are set up behind the inspection chambers. They must be level and their height must coincide with any datum levels mentioned on the plan.

(2) Pegs are inserted to show the trench width, and the trenches are excavated by mechanical digger where the amount of work justifies their use. The last 75 mm of spoil should be got out by hand immediately before the bedding is placed. Short, shallow trenches can be excavated by hand and in each case excavating should start from the lower end, where a sump or temporary drain can be provided to prevent trenches becoming muddy in wet weather.

(3) Timbering should be carried out as necessary (see Volume 1). Two further methods are shown in figures 5.39 and 5.40.

(4) Cast concrete bases for the inspection chambers; fix the channels and one pipe pointing from each channel in the correct direction.

(5) Attach a taut line and lay the pipes; the barrels must rest on the ground or in the concrete beddings and not on their collars.

(6) For rigid-jointed clay pipes, a strand of gaskin is wrapped round each spigot to centre the pipe in the collar of the previous pipe and to prevent any collaring mortar being forced through into the barrel of the pipe. A badger can be used to check this (figure 5.41). The joints should be caulked up with cement and sand in the proportion 1:2 and flaunched off at 45° (gaskin/cement joint, figure 5.19). The Building Regulations recommend the use of flexible joints for rigid pipes.

(7) Every fifth pipe should be boned in (figures 5.42 and 5.58), collars should be protected against the elements with sacking to prevent premature drying out.

(8) Notify the local authority before haunching or covering a drain (24 hours' notice is required).

(9) Cover the drain as required; no large stones are be used in the first 300 mm of backfill and tamping is to be light up to this point.

(10) Send notice to the local authority not more than 7 days after the completion of backfilling.

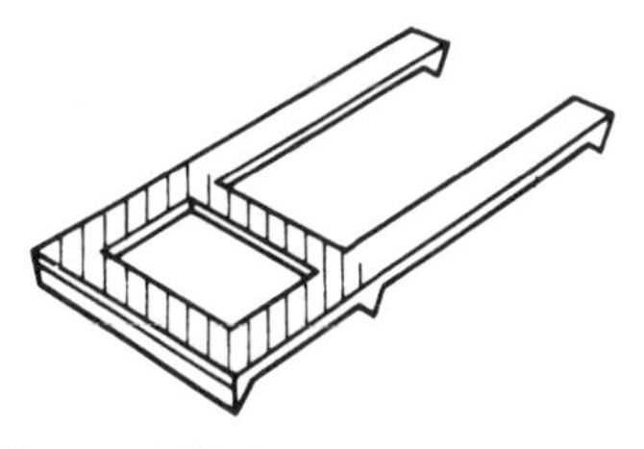

Figure 5.38 Galvanised step iron

TESTING DRAINS

The Building Regulations direct that any drain or private sewer shall on completion of the works, including backfilling, etc., be capable of withstanding a test for watertightness. Testing should be carried out from inspection chamber to inspection chamber, including any short branches; long branches should be tested separately. The length of drain between the

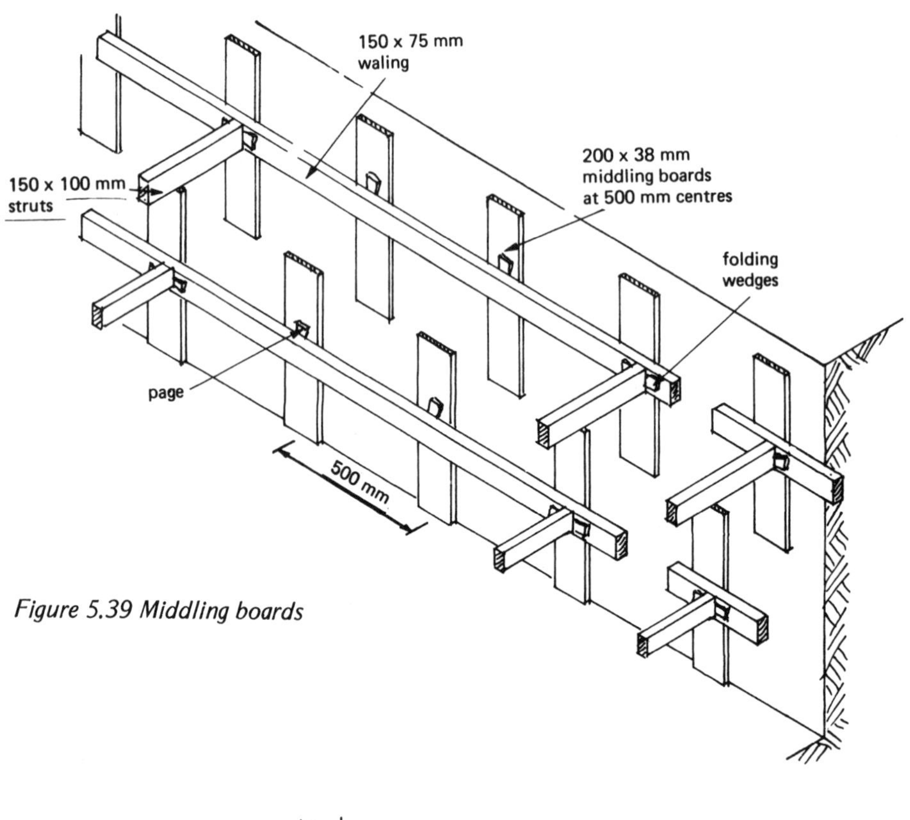

Figure 5.39 Middling boards

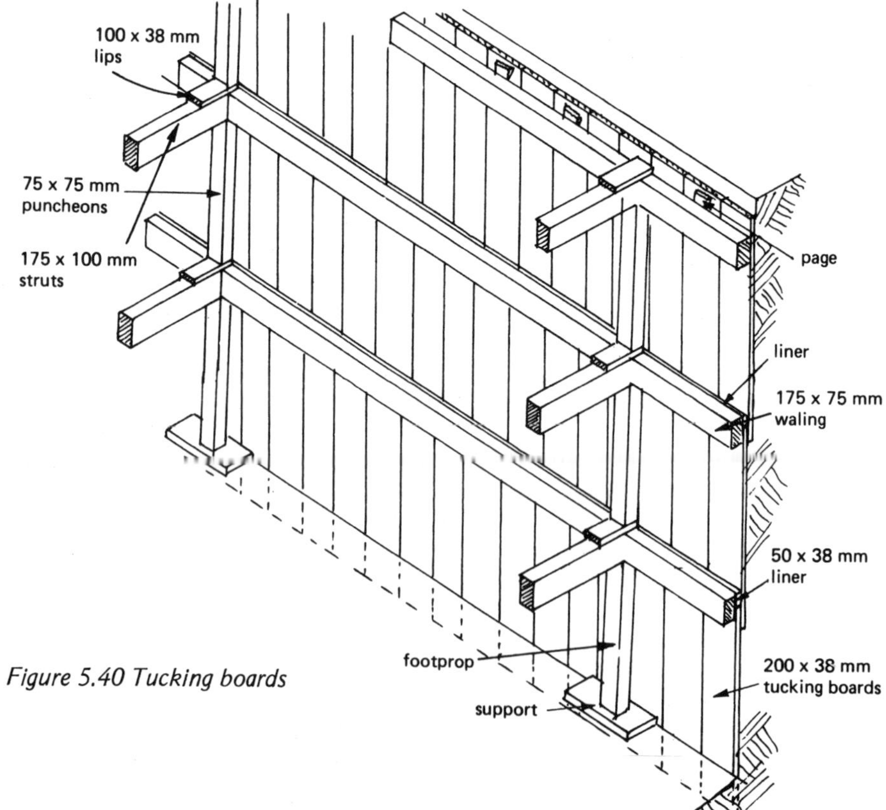

Figure 5.40 Tucking boards

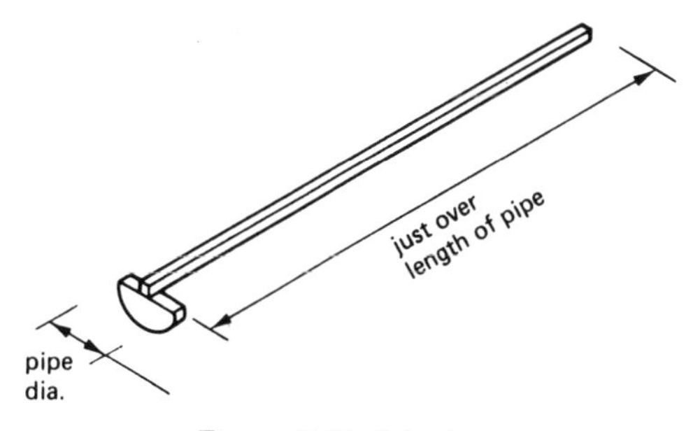

Figure 5.41 A badger

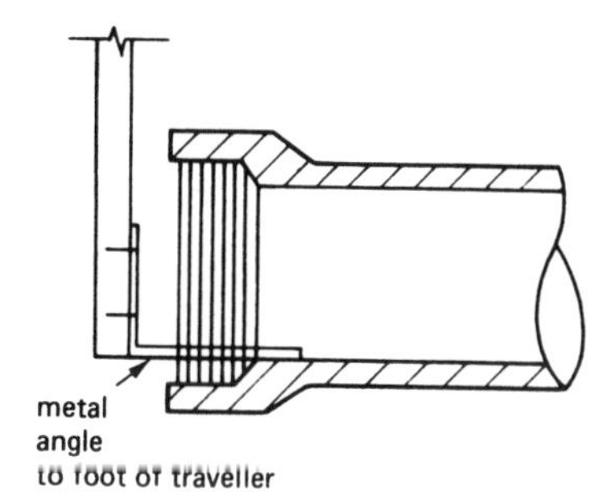

Figure 5.42 Detail of foot of traveller

last inspection chamber on site and a saddle on the main sewer should be tested via a testing junction installed close to the main sewer, which is sealed off before backfilling takes place.

The Water Test (figure 5.43)

This is the most widely used test and is generally considered to be the most reliable. Where rigid joints have been used it is important that at least 24 hours should elapse before testing, to allow the mortar to gain sufficient strength, but flexibly jointed pipes can be tested immediately.

The test is applied by plugging the lower end of the pipeline with either an expanding rubber ring plug or an air bag stopper (figures 5.44 and 5.45), and filling the pipeline with water to provide a head of 1.5 m above the higher end. It is important not to subject the lower end to more than a 6 m head to avoid over-stressing the joints. The section of drain should be

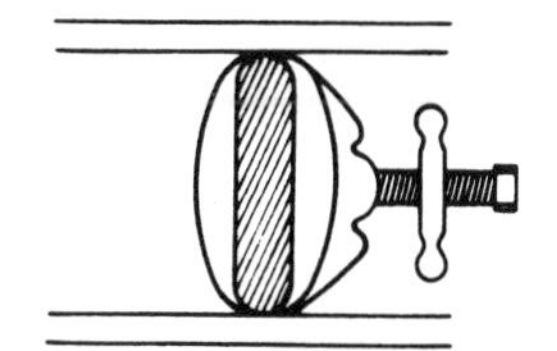
Figure 5.44 Expanding drain plug

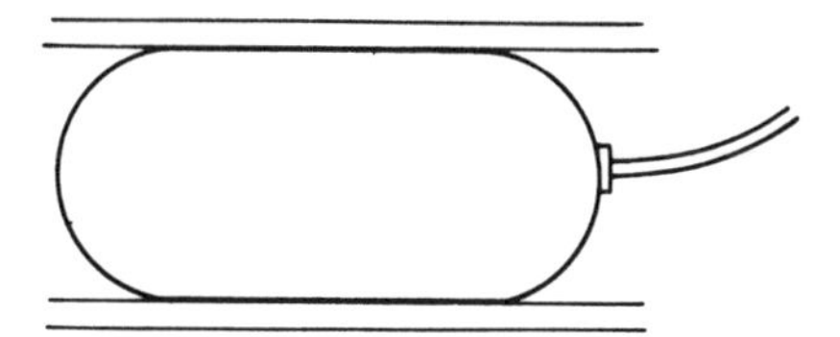
Figure 5.45 Inflated air-bag stopper

filled, left to stand for 2 hours and then topped up. The leakage over 30 minutes should be measured and should not be more than 0.05 litres for each metre run of 100 mm drain - a drop of 6.4 mm/m; and 0.08 litres for a 150 mm drain - a drop of 4.5 mm/m.

Note Where this test is being carried out in water-logged trenches, colouring powder should be placed in the testing water, and any leaks will then be quickly noticed.

The Air Test (figures 5.46 and 5.47)

Where water is not available or its disposal is inconvenient, this test is considered to be a good alternative.

The test is carried out by firmly plugging each end of the pipeline and pumping in air until a pressure of 100 mm is indicated on the manometer. The pressure should not fall from 100 mm to below 75 mm during a period of 5 minutes. If the pressure does fall below this a leak is indicated, and if this cannot be located, a smoke bomb can be used.

The Smoke Test

A smoke bomb or smoke-generating machine is used to supply smoke at the lower end of the system. The top of the vent pipe is plugged and the seals in the gullies are removed until smoke is seen emerging. The

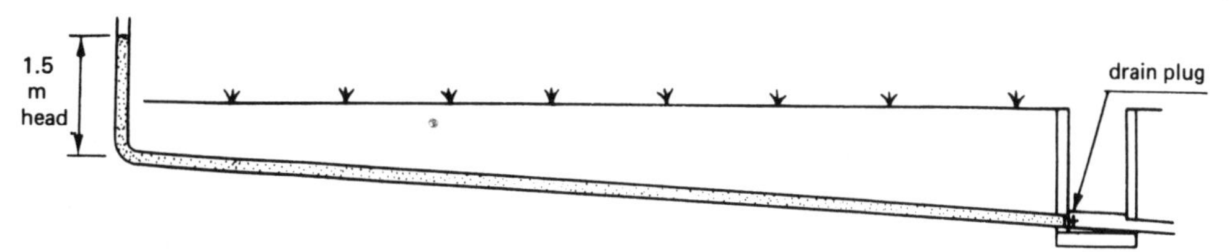

Figure 5.43

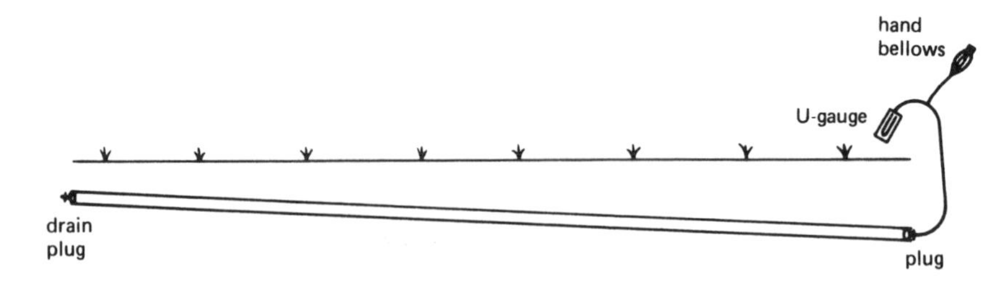

Figure 5.46

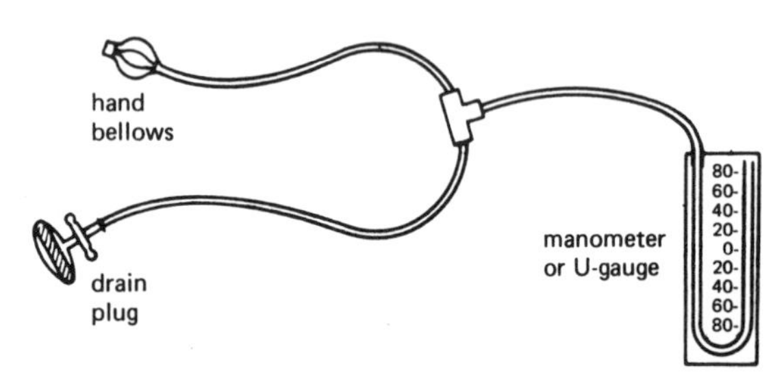

Figure 5.47

water seals are replaced and the seal at the top of the vent pipe is removed until smoke emerges. The seal is reinstated and smoke continues to enter the pipeline. This test is considered to be imprecise and is not recommended.

While the three tests mentioned will show any leaks that exist, there may be internal obstructions within the pipeline that will cause a blockage in the course of time (figure 5.48). This can be decided in two ways.

The Ball Test

A smooth, solid rubber ball, 13 mm less in diameter than the bore of the pipe, is inserted into the top end and should roll freely down the invert of the pipe. If it stops, a blockage is indicated. This is located by inserting drain rods into the pipeline until they touch the ball; the rods are then removed and laid alongside the drain to show the position of the blockage, which should be corrected as necessary.

The Reflection Test

A lamp and a mirror should be placed in the inverts of adjacent inspection chambers, as shown in figure 5.49. Light is reflected along the drain and the condition of the bore can be examined. This test can only be used where the drain is perfectly straight.

REPAIRS AND ALTERATIONS TO DRAINS

Inserting a New Pipe or Junction in a Straight Run

Occasionally a pipe may be damaged, possibly as a result of excavations, or a new set of pipes has to be connected into an existing run. There are two usual methods of carrying out this work.

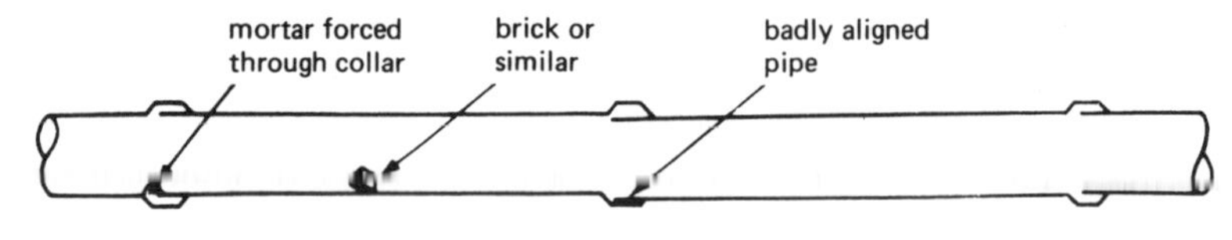

Figure 5.48

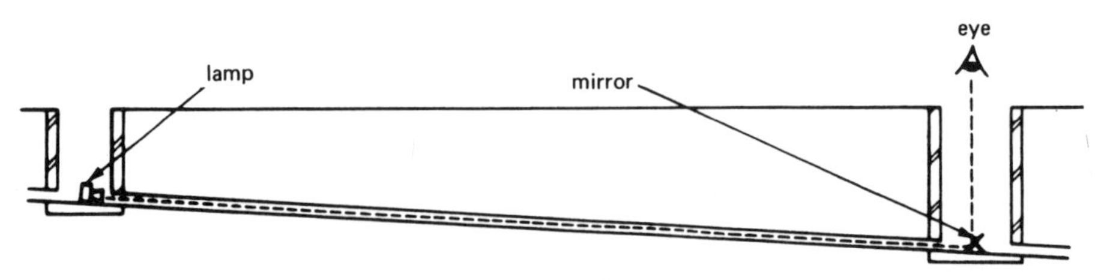

Figure 5.49

Method 1

(1) Expose the broken pipe and one more pipe on either side of it (figure 5.50*a*).
(2) Break out the three pipes, taking great care not to damage the collar and spigot on either side (figure 5.50*b*).
(3) Insert the new pipes as shown in figure 5.50*c*. They will drop into position and can be lined up and jointed (figure 5.50*d*).

Method 2

(1) Expose both the pipe that is to be replaced with a junction and the pipe above it in the run (figure 5.51*a*).
(2) Carefully break out the two pipes (figure 5.51*b*).
(3) Slide a loose collar on to the spigot remaining and place in position the junction and a butt pipe (figure 5.51*c*).
(4) Slide the loose collar over the joint, line up and make good all joints (figure 5.51*d*).

Connections to Inspection Chambers

(1) Cut away the benching on the appropriate side.
(2) Cut a suitably sized hole in the side wall.
(3) Bed the required splayed-end channel bend in position and connect the new pipe run to this.
(4) Make good to the walls and re-form the benching.

Connecting to a Main Sewer using a Saddle

(1) Check the depth of flow in the main at a convenient inspection chamber and cut a hole in the cheek above the flow to avoid surcharging the branch (figure 5.52). This can be carried out with a small hammer and a sharp chisel.

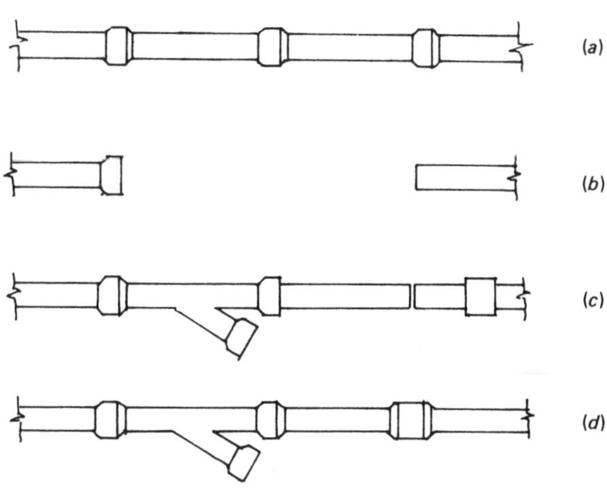

Figure 5.51

(2) Carefully enlarge the hole with the same tools and/or a scutch, cutting at the thickness of the main rather than on the face.
(3) When the spigot of the saddle fits snugly, clean any pipe debris out of the main and bed the saddle in cement mortar.
(4) Putting your hand through the saddle, clear away any mortar that has squeezed through, connect the pipework and surround it in concrete.

Note The local authority will normally inspect the connection before concreting is carried out and some authorities require the insertion of a testing junction close to the sewer connection in order to test the complete system. After testing, the junction is sealed off.

Cutting Drainpipes

Apart from the use of a special cutting tool, the following two methods are in common use.

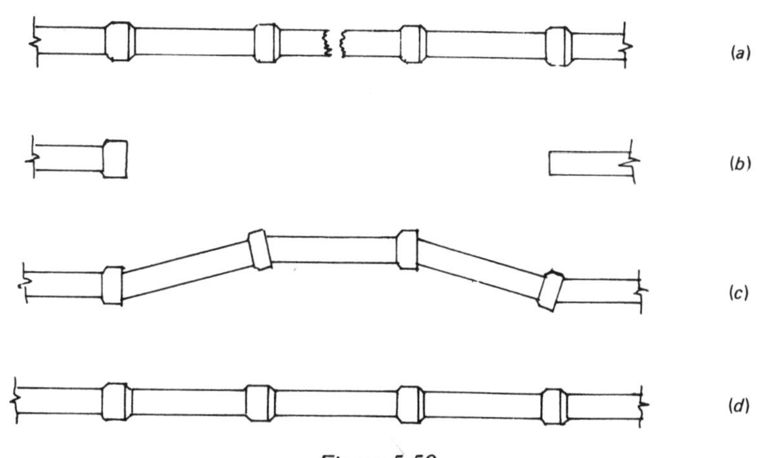

Figure 5.50

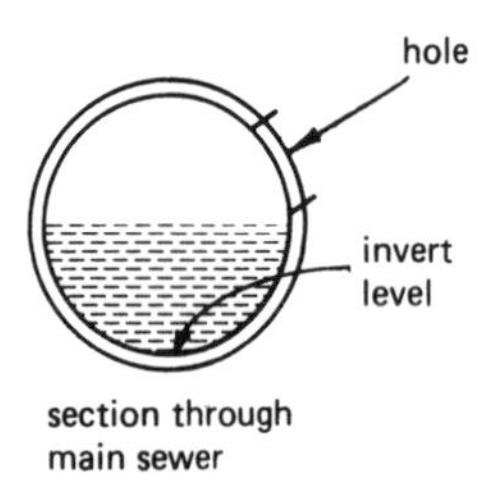

Figure 5.52

(1) Stand the pipe on end on a reasonably soft surface and fill with sand up to the required cutting mark (figure 5.53). Tap round the pipe on the cutting mark with a small hammer and sharp chisel, or a scutch, until fracture occurs.
(2) Form a mound of sand and with the barrel of the pipe resting on it at the cutting mark and with the spigot unsupported, tap as before while rotating the pipe (figure 5.54).

Note Where channels are required but only straight pipes or bends are available, splitting is possible by carefully tapping the pipe along each side in turn until fracture occurs.

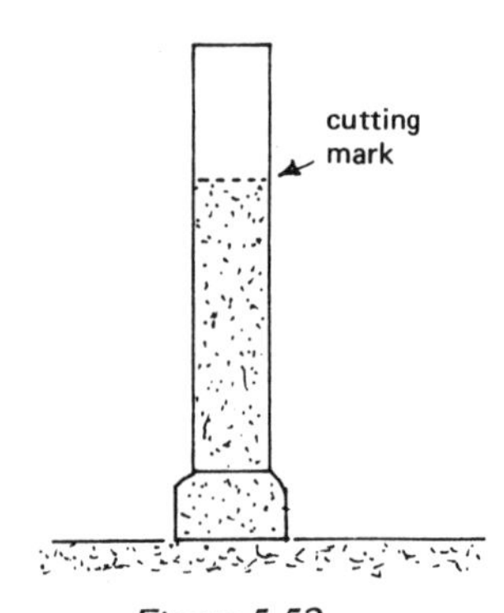

Figure 5.53

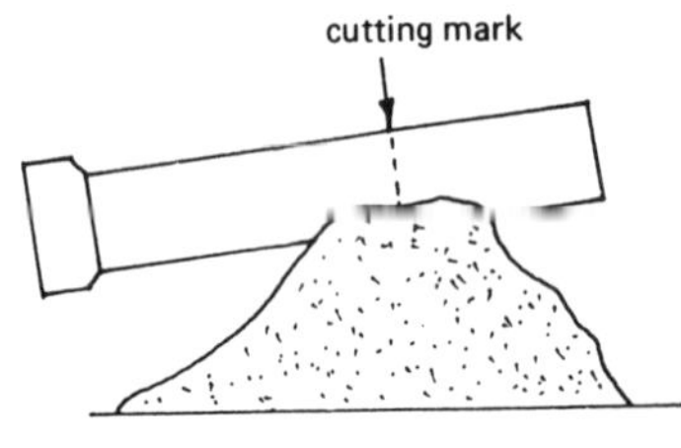

Figure 5.54

CALCULATION OF INVERT LEVELS

Before any pipes are laid it is necessary to know the invert level of the drain or sewer to which the connection is to be made and to relate this to the ground level point at which pipelaying is to start. The invert level is arrived at in one of three ways

(1) the local authority may be able to provide the information
(2) by calculation
(3) by excavation vertically downwards to expose the main

If the local authority is unable to provide the information and excavating is inconvenient, calculation is as follows.

(1) Remove the inspection chamber covers on either side of the proposed saddle position and with a Cowley or tilting level obtain the difference between the invert levels (figure 5.55).
(2) Measure the overall distance between these inspection chambers and the distance from each to the saddle (figure 5.56).
(3) Calculate the invert level of the drain at the saddle as follows. Referring to figures 5.55 and 5.56, since

$$\text{total fall} = 1 \text{ m } (2.6 - 1.6)$$

the main is laid at a fall of 1 in 30 (the distance between inspection chambers is 30 m). To find the fall in 18 m

$$\frac{1}{30} = \frac{\text{fall}}{18}$$

Cross multiplying

$$30 \times \text{fall} = 1 \times 18$$

dividing both sides by 30

$$\frac{30 \times \text{fall}}{30} = \frac{1 \times 18}{30}$$

Cancelling

$$\text{fall} = \frac{3}{5} = 0.6 \text{ m}$$

Thus the fall from inspection chamber B to the saddle is 0.6 m.

Note It may be simpler at this stage for the student to remember that

$$\text{fall} = \frac{\text{actual distance}}{\text{given distance}} = \frac{18}{30} = 0.6 \text{ m}$$

Since

$$\text{staff reading at B} = 1.6 \text{ m}$$

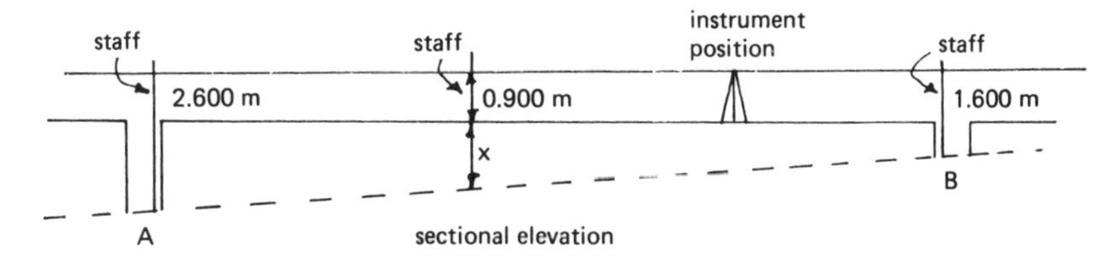

Figure 5.55

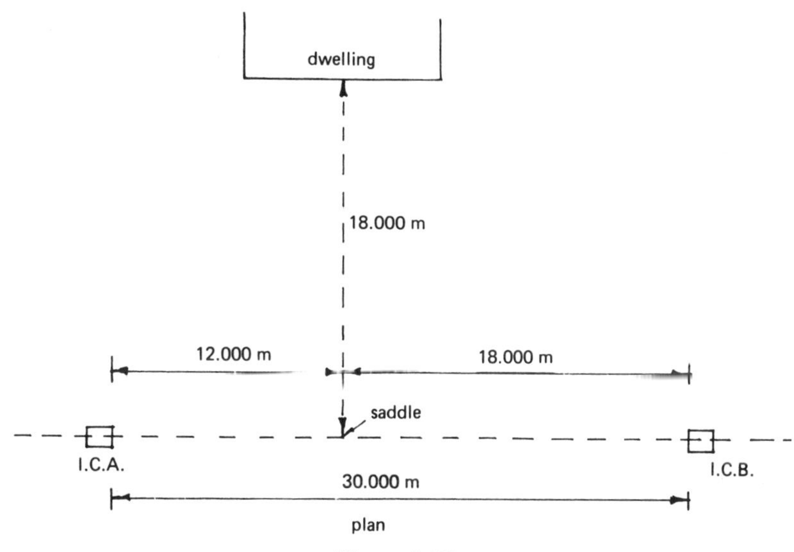

Figure 5.56

depth of invert at saddle = 1.6 + 0.6 (depth at B + fall)
= 2.2 m

And since

ground level staff reading at the saddle = 0.9
depth of invert below ground level = 2.2 – 0.9
= 1.3 m

That is

$$x = 1.3 \text{ m}$$

Having found this it is necessary to relate it to the invert level at the topmost inspection chamber of the proposed drain to ascertain that the fall will be suitable. Assume that the invert level is 0.6 m below ground level (see figure 5.57). An instrument is set up between these points and readings are taken of 1.600 and 1.4 as shown. Since the invert level at X is 0.6 below ground level

fall from invert level at X to invert Y
= 0.5 m (2.7 – 2.2)

Thus the fall is 1 in 36, which is quite satisfactory (0.5 in 18 m equals 1 in 36). Sight rails can now be set up at each end (figure 5.58), and the ground is excavated. Boning rods are used to obtain the required slope and the pipes are laid.

Note If, when the depth of the inspection chambers is checked in the first place, it is obvious that there will be an adequate fall, the above procedures may be unnecessary.

While the foregoing explanations on invert levels were necessary, it is probable that all that the craft student will be required to do is to calculate the total fall between inspection chambers, given the overall length and required fall.

Example 5.4

A drain is to be laid a distance of 45 m in level ground and the fall is to be at 1 in 60. Calculate the total fall (figure 5.59).

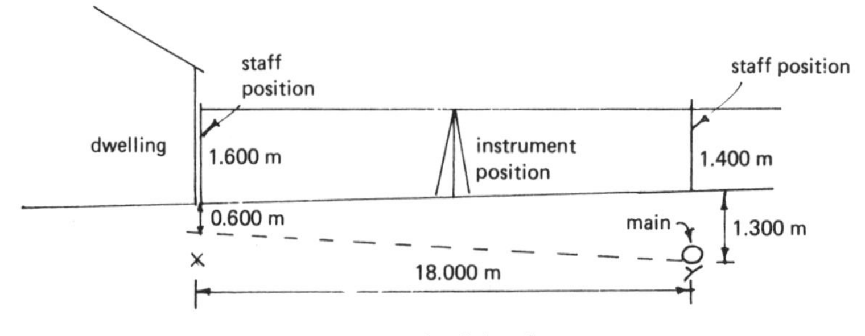

Figure 5.57

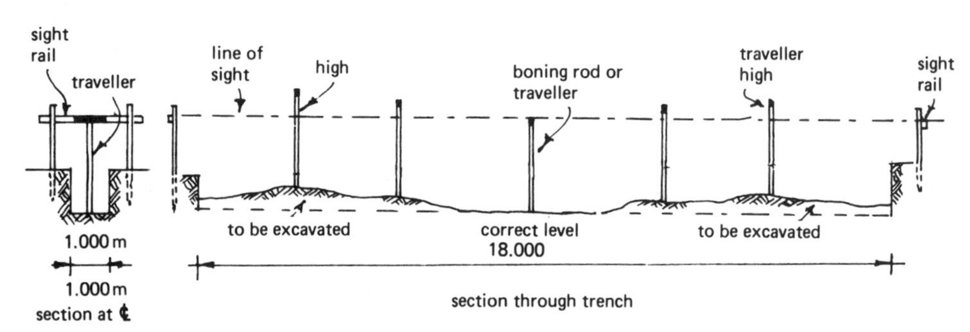

Figure 5.58

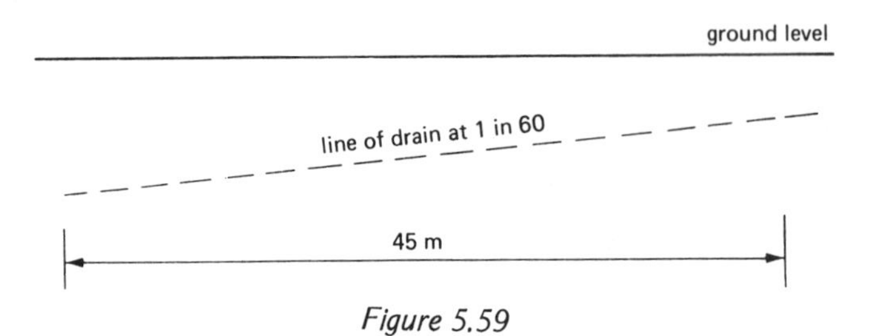

Figure 5.59

$$\text{Total fall} = \frac{\text{actual distance}}{\text{given distance}}$$

$$= \frac{45}{60}$$

$$= 0.75 \text{ m}$$

$$= 750 \text{ mm}$$

Possibly the invert level may have to be found given the level at one end, the overall distance and the fall.

Example 5.5

The invert level at the top end of a length of drain 50 m long is 2.50 above datum and the fall has to be 1 in 40. Find the invert level of the bottom inspection chamber (figure 5.60).

$$\text{Total fall} = \frac{\text{actual distance}}{\text{given distance}}$$

$$= \frac{50}{40}$$

$$= 1.25 \text{ m}$$

Since

invert level at top end = 2.50

then

invert level at bottom = 2.50 – 1.25

= 1.25 A.D. (above datum)

Use of a Tapered Straightedge

Another method of obtaining the correct fall is to use a tapered straightedge in conjunction with a spirit

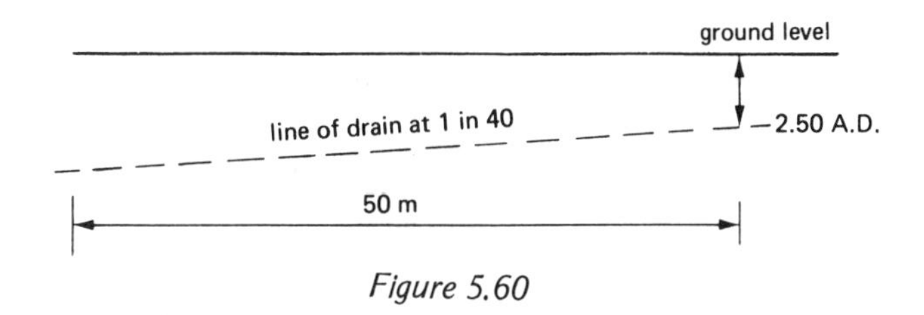

Figure 5.60

level. For example, if a fall of 1 in 40 is required, a 4 m straightedge is ideal. This should be about 150 x 25 in section and it is cut down lengthways as shown in figure 5.61. That is

$$\text{fall} = \frac{\text{actual distance}}{\text{given distance}}$$

$$= \frac{4}{40} \frac{(4\text{ m straightedge})}{(1\text{ in }40)}$$

$$= \frac{1}{10}\text{ m}$$

$$= 100\text{ m}$$

Therefore, the straightedge should taper by 100 mm, say from 150 to 50 mm, in 4 m, and a spirit level should be placed on top centre to check the fall (figure 5.61).

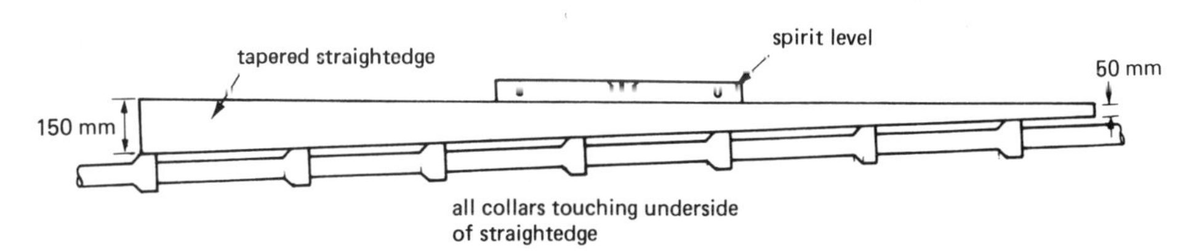

Figure 5.61

6
SCAFFOLDING

CONSTRUCTION REGULATIONS

Where men are unable to reach their work from the ground or part of a building, ladders or a scaffold must be provided. Tubular scaffolding is a temporary structure, erected to support a platform or number of platforms, at different heights, and may be of steel or aluminium alloy. Its erection is strictly governed by the Construction (Working Places) Regulations, which came into operation in August 1966 and have not as yet been metricated; therefore any measurements given have been converted.

It is the duty of every employer and employee to comply with the requirements of the Construction Regulations, which are summarised as follows.

(1) A sufficient quantity of materials is to be provided, to be sound and of adequate strength for its purpose.
(2) The erection, alteration and dismantling of a scaffold must be carried out under the supervision of a competent person.
(3) All materials intended for a scaffold must be inspected by a competent person, who must also inspect the completed scaffold at least every 7 days and after exposure to adverse weather conditions. The results of an inspection must be entered in the prescribed register (figure 6.1).
(4) Any timber to be used must be in good condition, of suitable quality and not painted in such a way that defects are hidden.
(5) Scaffolds must not be overloaded and materials are not to be kept on a scaffold unless required within a reasonable time.
(6) Partly dismantled scaffolds must comply with the Construction Regulations or carry permanent warning notices. The access to incomplete scaffolds should be effectively blocked as far as possible.
(7) Loose materials such as bricks, drainpipes, chimney pots, etc. must not be used as supports for platforms, but a firm packing of bricks or blocks may be used if stable up to a height of 600 mm above ground level.
(8) A platform must extend beyond the end of a wall at least 600 mm if work is to be carried out at that point.

SCHEDULE — Regulation 22

FACTORIES ACT 1961

CONSTRUCTION (WORKING PLACES) REGULATIONS 1966

SCAFFOLD INSPECTIONS

FORM OF REPORTS OF RESULTS OF INSPECTIONS UNDER REGULATION 22 OF SCAFFOLDS, INCLUDING BOATSWAIN'S CHAIRS, CAGES, SKIPS AND SIMILAR PLANT OR EQUIPMENT (AND PLANT OR EQUIPMENT USED FOR THE PURPOSES THEREOF)

Name or title of Employer or Contractor . . . .

Address of Site . . . .

Work Commenced–Date . . . .

Location and Description of Scaffold, etc. and other Plant or Equipment Inspected (1)	Date of Inspection (2)	Result of Inspection. State whether in good order (3)	Signature (or, in case where signature is not legally required, name) of person who made the inspection (4)

Figure 6.1

TUBES, FITTINGS AND BOARDS

Members (Tubes)

Standards

Standards are the upright members of a scaffold and they are usually spaced between 1.8 and 2.5 m apart, depending on the load to be carried and the type of work being done. They must be vertical or slightly inclined towards the building and sufficiently close to ensure stability. A firm base is essential and they can be extended where required, using joint pins or sleeve couplers, the height of which should be staggered.

Ledgers

Ledgers are long, horizontal members, which are fastened to the standards on the inside using load-bearing couplers. They are normally secured together lengthways with sleeve couplers or joint pins, and if the latter are used they must be positioned at one-third of the bay owing to their lack of tensile strength.

Putlogs

Putlogs are short, flat-ended tubes 1.2 to 1.5 m long, which are inserted into the bed joints of brickwork to the full extent of the flat supporting surface. They are usually fastened to the ledgers with putlog couplers and in each bay one putlog must be within 300 mm of a standard. The spacing of the putlogs varies according to the thickness of the planks.

Plank thickness (mm)	32	38	50
Putlog spacing (m)	1	1.5	2.5

Transoms

Transoms are short lengths of tube that take the place of putlogs in an independent scaffold, both ends being supported by ledgers to which they are secured using putlog couplers.

Longitudinal Braces

Longitudinal braces are lengths of tube fastened at or as near to 45° as possible on the outside of standards to provide stability and eliminate sideways movement. They are required every 30 m and must extend to the full height of the scaffold. It is preferable to fasten them to putlogs or transoms with double couplers, or alternatively to standards with swivel couplers.

Cross Braces

Cross braces are short lengths of tube used to connect and give added rigidity to alternate pairs of standards in an independent scaffold. They can be fastened to ledgers with double couplers or to standards with swivel couplers.

Puncheon

A puncheon is a vertical, loadbearing member, not taking its support from the ground. It is used, for example, where an opening has to be formed through a scaffold for a lorry entrance/exit. Standards are placed either side of the opening with puncheons between. Extra braces, known as spurs, are needed to strengthen the structure. Figure 6.4 shows a line diagram to explain the use of spuncheons/spurs.

Bridle Tube

A bridle tube is a horizontal tube secured just clear of the wall face in a putlog scaffold. It is secured across openings below the putlogs on either side with double couplers and is used to support extra transoms as required to carry the boards (figure 6.2). Where a wide opening occurs the centre transom can be supported off the window bottom if necessary.

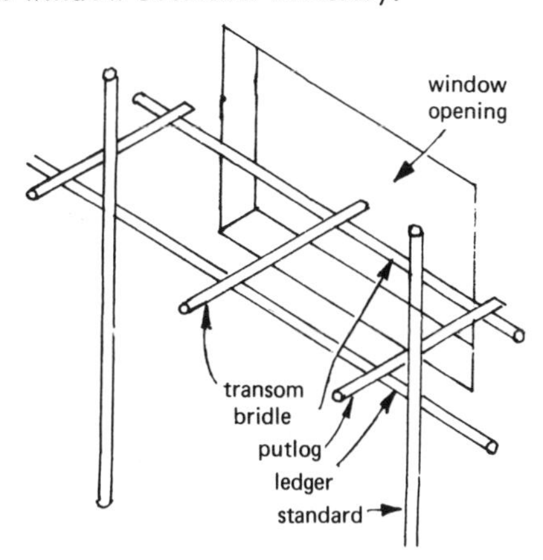

Figure 6.2 Bridling an opening

Guard Rails (figure 6.3)

Guard rails are lengths of tube which must be provided where men are liable to fall more than 2 m. They must be secured on the inside of the standards at a height of between 0.9 m and 1.125 m and kept permanently in position except for access and loading.

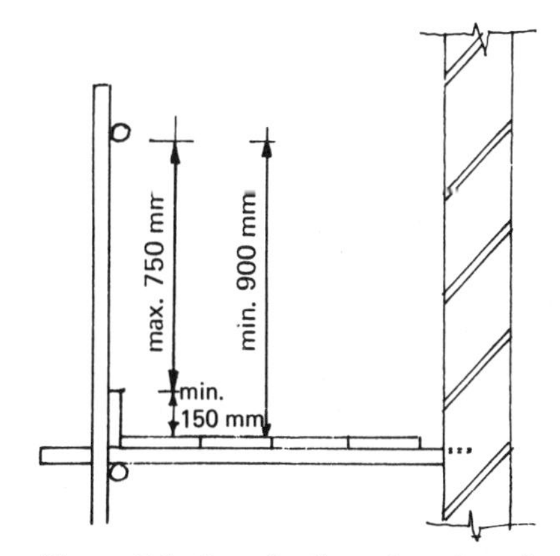

Figure 6.3 Guard rails and toe boards

Raking Tube or Raker

This is a length of tube which can be used to prop up a scaffold initially before the insertion of reveal or through ties. Where a wall contains no openings for ties, rakers provide an alternative method of preventing a scaffold from pulling away from the wall. They should be fixed at 45° or as near as possible using double or swivel couplers and be provided with a sound foot block at the base (figure 6.25).

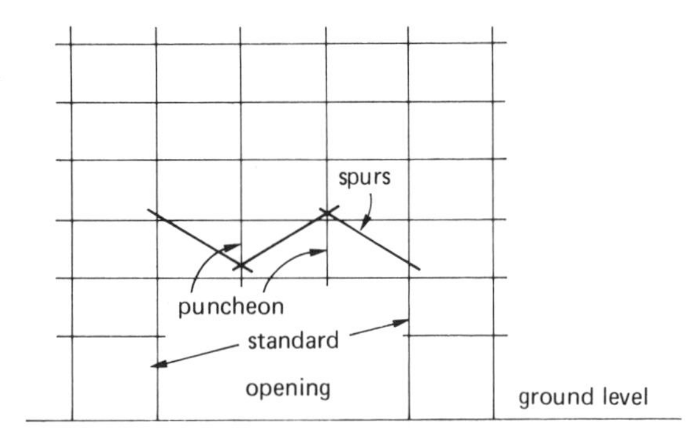

Figure 6.4

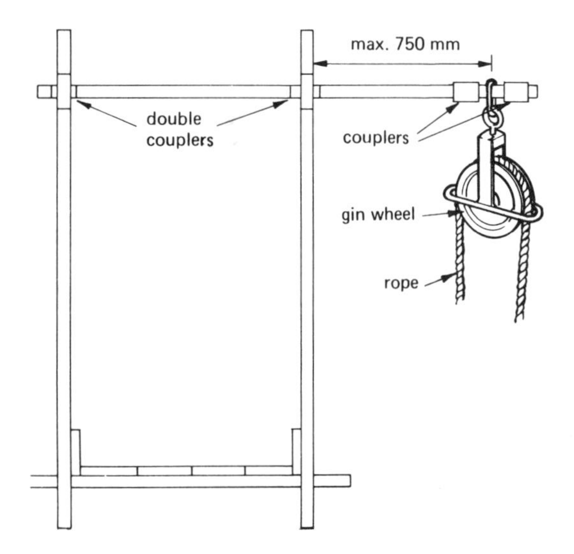

Figure 6.5 Drop-forged double coupler

Butting Pieces

These are very short lengths of tube used, for example, to reinforce across a joint pin. A butting piece should be securely fastened on either side with parallel or universal couplers.

Scaffold Fittings

Many types of fitting are produced by different firms; those shown in figures 6.6–6.21 are S.G.B. scaffold fittings.

Double Couplers (figure 6.6)

These are used for fastening ledgers to standards and in all positions where strength is required, for example, bridle tubes to putlogs.

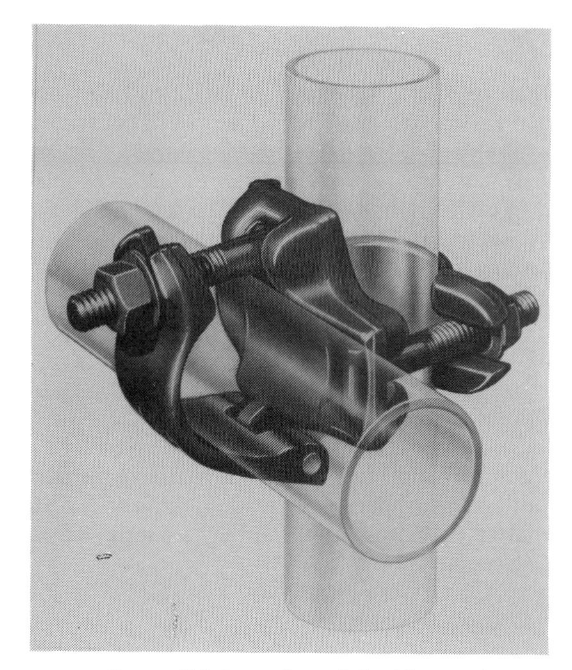

Figure 6.6 Drop-forged double coupler

Universal Coupler (figure 6.7)

This is another 90° coupler, which can also be used for connecting two loadbearing tubes in parallel.

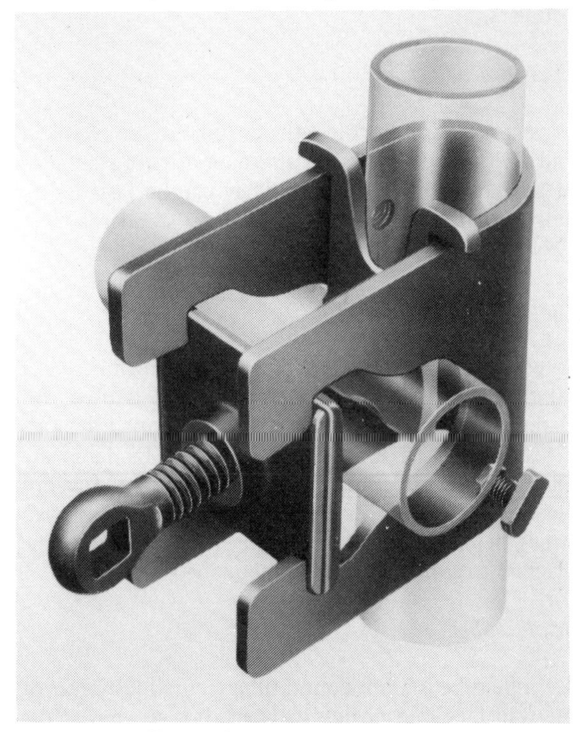

Figure 6.7 Universal coupler

Putlog Couplers (figures 6.8 and 6.9)

Used for connecting putlogs or transoms to ledgers.

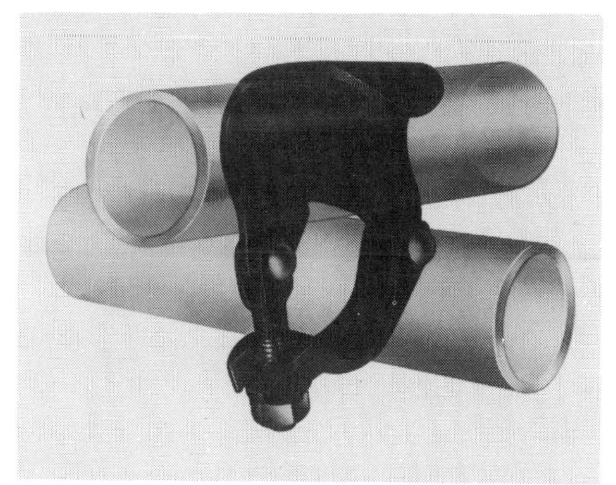

Figure 6.8 Drop-forged putlog coupler

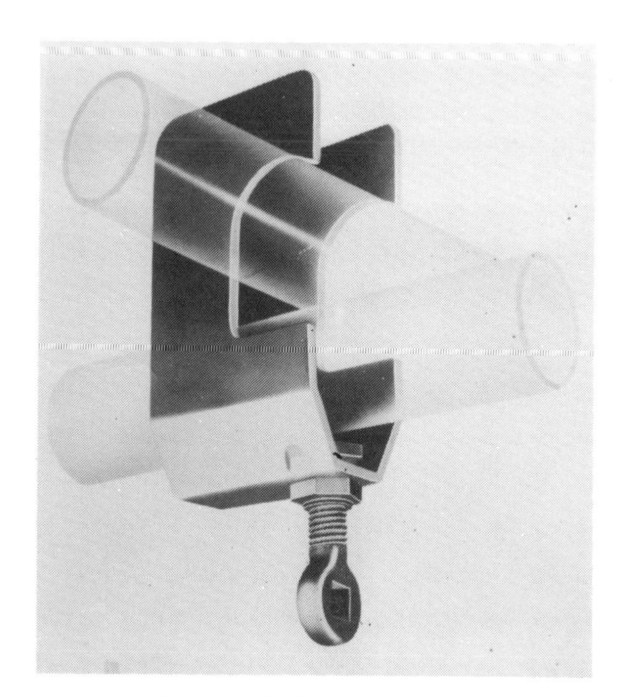

Figure 6.9 Putlog or brace coupler

Swivel Coupler (figure 6.10)

This is a one-piece coupler used for connecting two scaffold tubes at any angle through 360°, for example, longitudinal or cross braces to standards.

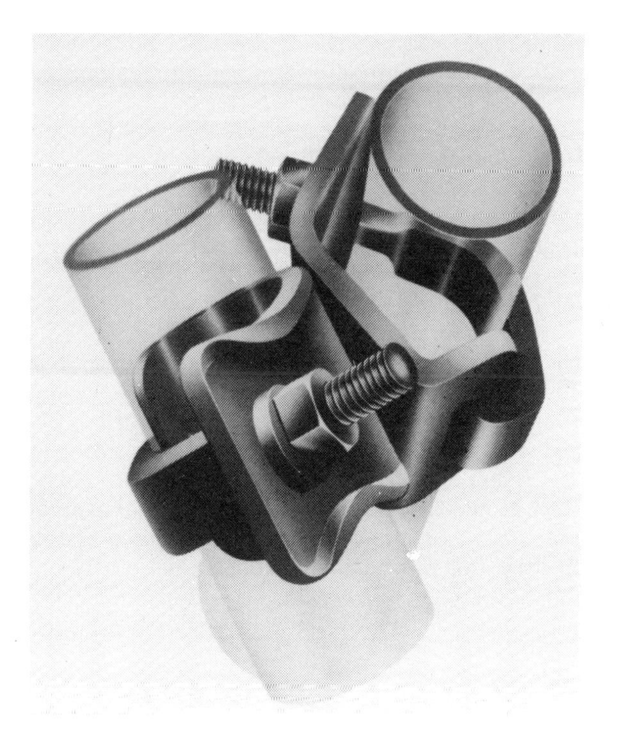

Figure 6.10 Swivel coupler

Putlog End (figure 6.11)

This is a simple fitting which will convert a transom into a putlog. When a putlog scaffold is being erected against an existing brick building it is easier to cut the

Figure 6.11 Putlog end

putlog holes where required, insert the putlog ends secured with hardwood wedges and fit transoms to the putlog ends as the scaffold is erected.

Sleeve Coupler (figure 6.12)

This is an external fitting used to join standards, ledgers, longitudinal braces and guard rails end to end.

Joint Pin (figure 6.13)

The uses of a joint pin are as for a sleeve coupler, but it is not used for braces. It fits internally into the end of a scaffold tube and expands against the wall of the tube as the bolt is turned. They are not as strong as sleeves and must be used at one-third the bay width, never at mid-bay. If a joint pin should occur at mid-bay it must be reinforced with a butting piece, secured on either side with a universal coupler.

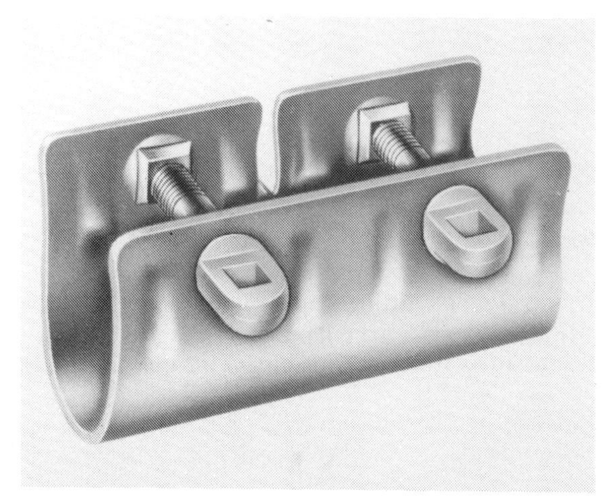

Figure 6.12 Sleeve coupler

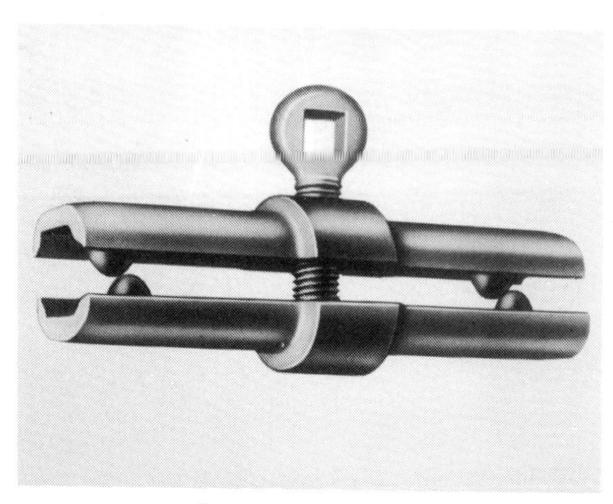

Figure 6.13 Joint pin

Base Plate (figure 6.14)

This is a 150 x 150 mm steel plate which is used to provide a flat, bearing surface for load distribution from standards. It has a central spigot 50 mm high,

Figure 6.14 Base plate

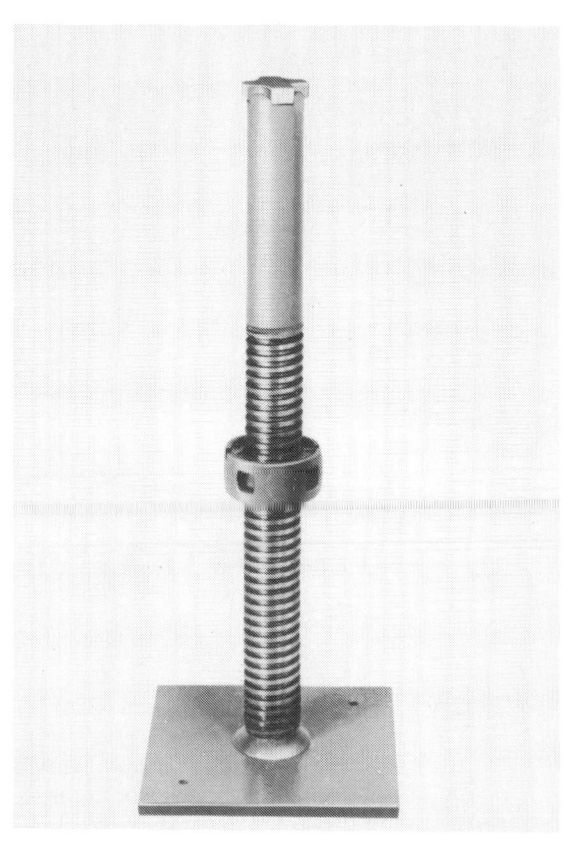

Figure 6.15 Adjustable base plate

on which the foot of the standard is located, and two fixing holes for use with sole plates.

Adjustable Base Plate (figure 6.15)

This is for use in undulating ground, particularly where settlement may take place. It has 230 mm of adjustment.

Reveal Pin (figure 6.16)

This fitting is inserted into the end of a short tube and is expanded as necessary to form a rigid horizontal or vertical tie in a window opening to which the scaffolding can be secured.

Figure 6.16 Reveal pin

Fixed Finial (figure 6.17)

This is used to connect a scaffold tube at right-angles to the extreme end of another tube without projection. It is very useful for guard rails, safety barriers, etc.

Toeboard Clip (figure 6.18)

This is used to secure a toeboard against a standard.

Gin Wheel (figure 6.19)

The figure shows a 250 mm steel wheel, with which a 19 mm-diameter rope is used. Its safe working load is 250 kg.

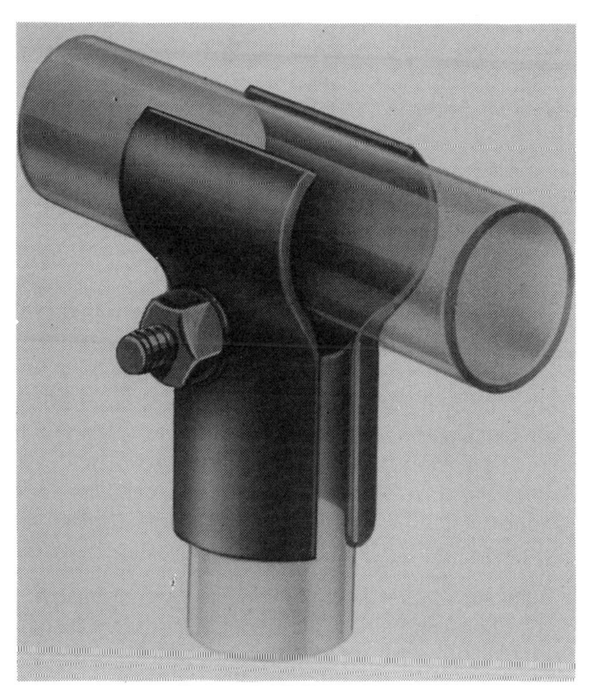

Figure 6.17 Fixed final

Figure 6.18 Toeboard clip

Castor Wheels (figure 6.20)

These are used for mobile scaffold towers and have foolproof wheel brakes, which cannot be accidentally released.

Spanners and Podgers (figure 6.21)

These are used for tightening and releasing the nuts on couplers.

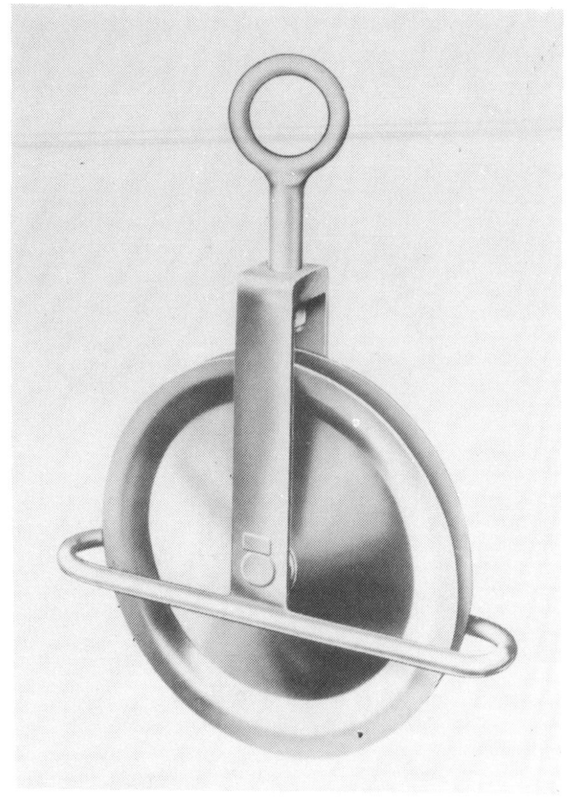

Figure 6.19 Gin wheel

Boards

Sole Plate (figure 6.22)

Where the bearing capacity of the ground surface is suspect sole plates should be placed beneath the base plates to further spread the load. A sole plate should support at least two standards.

Scaffold Boards

Scaffold boards must be at least 150 mm wide when 50 mm thick; 32 and 38 mm boards must be at least 200 mm wide. They must rest evenly and securely on their supports and each board should have at least three supports unless the span and thickness is sufficient to prevent sagging and make this requirement unnecessary. No board must project over its end support more than four times its thickness unless otherwise secured against tipping.

Toe Boards (figure 6.3)

Toe boards must be in position where men are liable to fall more than 2 m. Their purpose is to prevent men, materials, tools, etc. from accidentally falling from the scaffold. They are placed directly below the guard rails, and fixed to the standards with toe-board clips. Their minimum height is 150 mm and the maximum distance between the top of the toe board and the underside of the guard rail is 750 mm. As with guard rails they must not be removed except for access and loading.

Figure 6.20 Castor wheels

TYPES OF SCAFFOLD

The two main types of scaffold are putlog and independent scaffolding.

Putlog Scaffold (figures 6.22 and 6.23)

The putlog scaffold is mostly used where brick structures are being erected and is sometimes known as a bricklayer's scaffold. The scaffold depends for its

Figure 6.21 Scaffold spanners and podgers

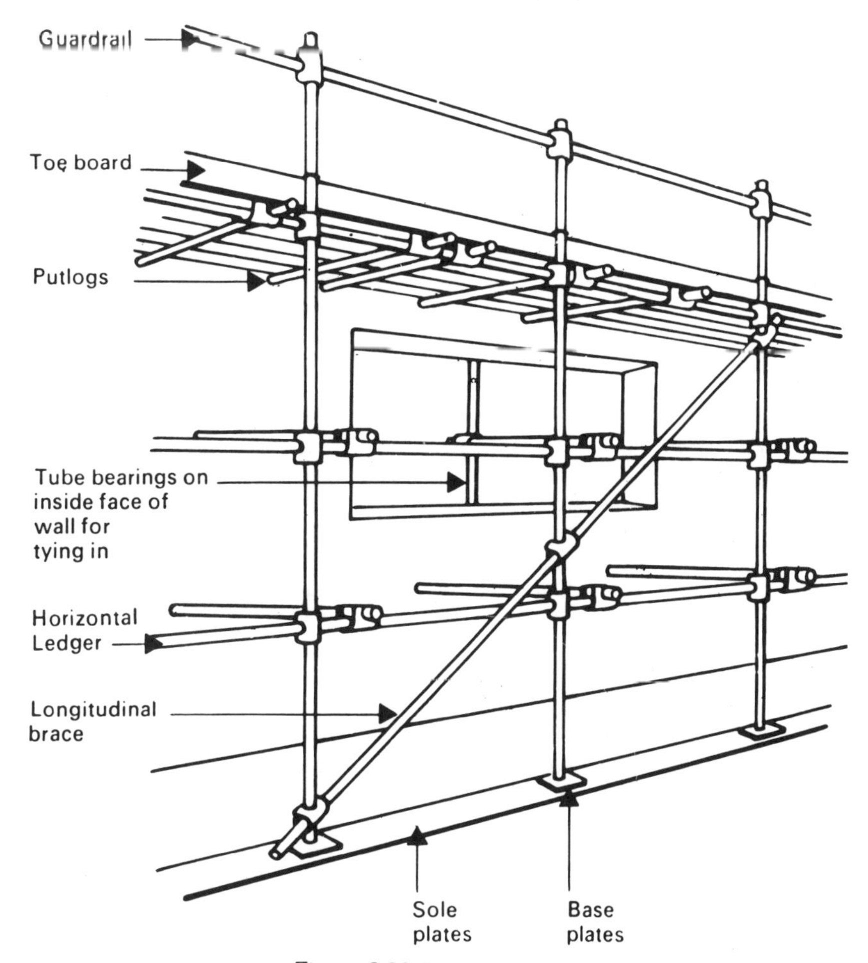

Figure 6.22 Putlog scaffold

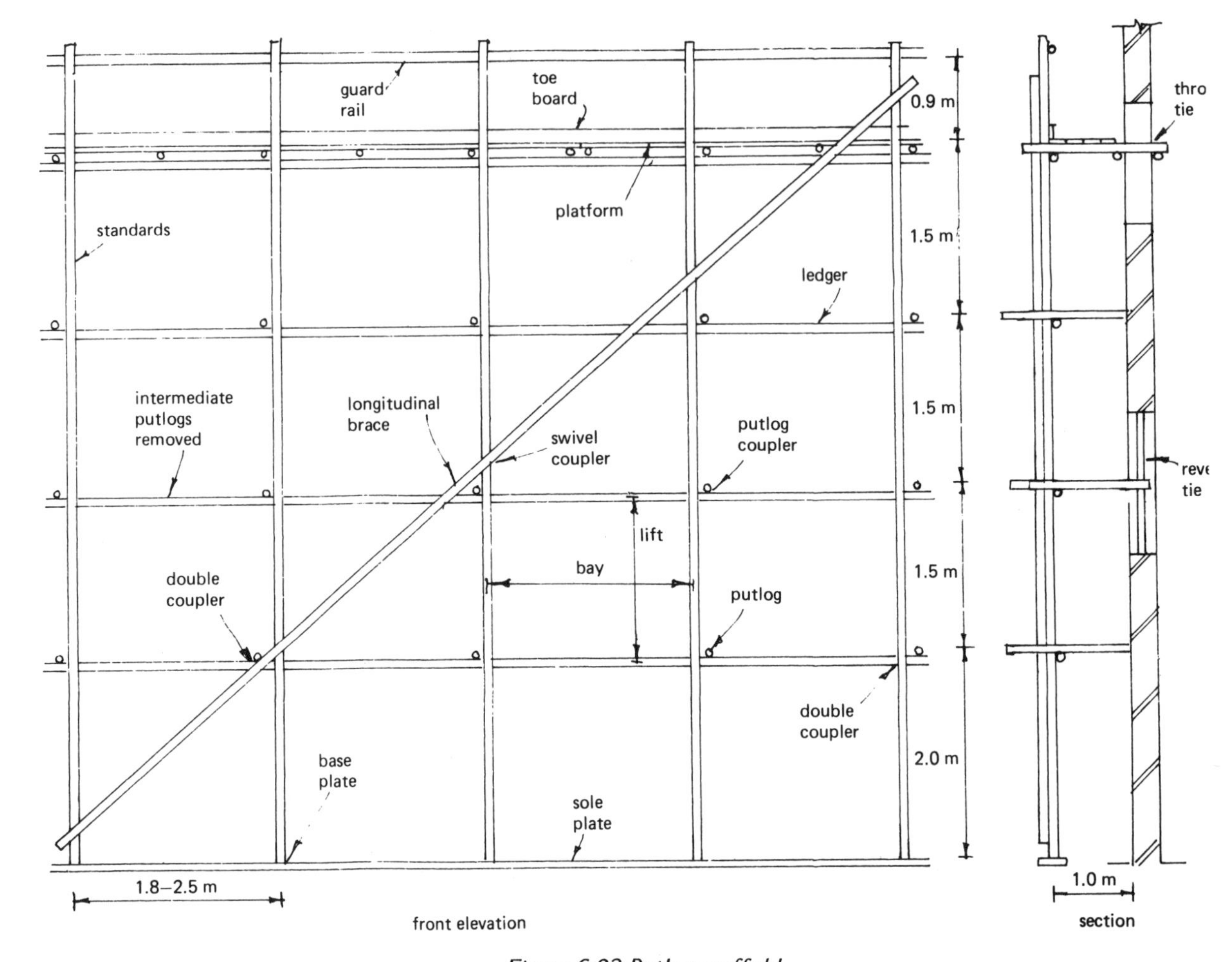

Figure 6.23 Putlog scaffold

support on the wall on the inside and the putlogs rest on ledgers which are supported by a single row of standards on the outside.

To erect a putlog scaffold a standard should be set up at each end or corner first and intermediate ones lined up from these. The space from the wall should allow for the required number of boards and a space at the wall face for a plumb level and for mortar droppings. The ledgers should be fixed to the standards with double couplers at approximately the height of the first lift, which is usually 2 m to allow for walking below. Subsequent lifts should be no more than 1.5 m in height. It must be remembered that coupling ledgers end to end with joint pins will provide little strength in tension and that sleeve couplers are, therefore, to be preferred for this purpose. As close to the standard as possible rest the flat end of a putlog on top of the first lift of brickwork, or wedge it into a raked-out bed joint, fixing the other end of the putlog to the ledger with a putlog coupler. If bricks or other materials are stacked on the platform as close to the standard as possible the stress in the ledgers will be reduced to a minimum.

Fix intermediate putlogs as required with due regard to the thickness of the planks, and plumb and level the scaffolding as it is erected, tightening fittings as work proceeds. When the length of the standards exceeds 6.5 m joint pins or sleeve couplers are used to connect the tubes together. As already mentioned, it is important not to have all the joints occurring at the same height, but to stagger them by using tubes of different lengths. With a putlog scaffold only one lift must be in use at a time.

Independent Scaffold (figures 6.24 and 6.25)

The independent scaffold is normally used on existing buildings or on structures where putlogs would be inconvenient. It is so called because it is self-supporting and carries all the superimposed loads without assist-

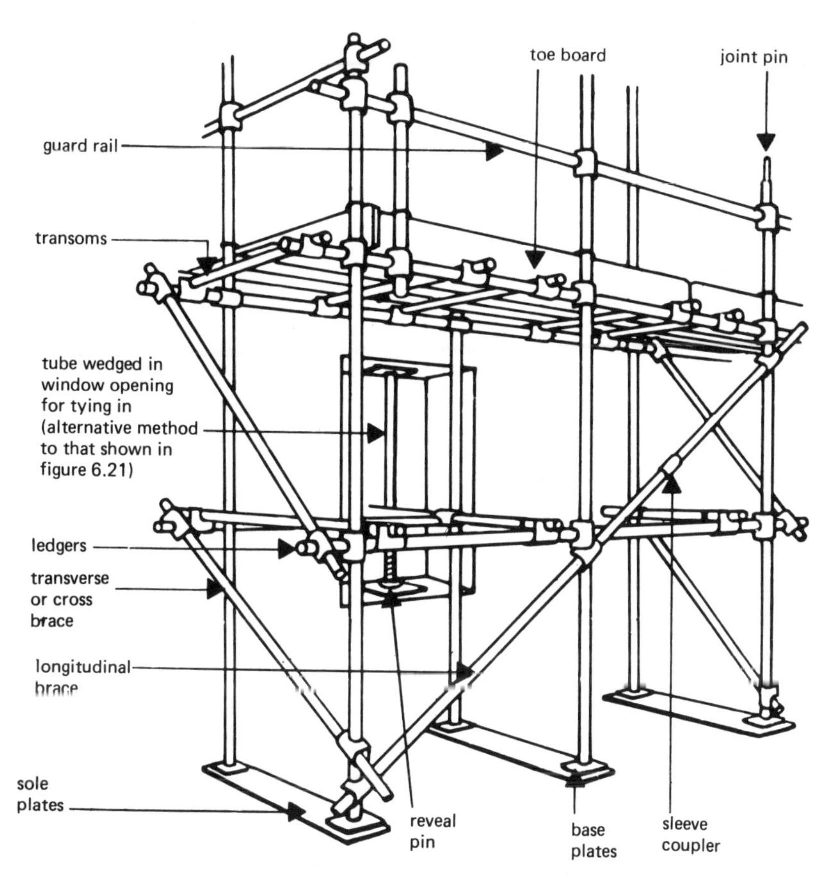

Figure 6.24 Independent tied scaffold

ance from the structure. It consists basically of two rows of standards, two rows of ledgers, transoms, longitudinal braces and cross braces.

When erecting this scaffold for bricklayers, set the inner row of standards about 330 mm from the wall so that the inside scaffold board can be placed on the transoms, projecting beyond the inner standards. Lifts are usually approximately 2 m. It is possible for one man to set up a simple independent scaffold using the 'mattress' method of erection as follows.

(1) Construct a temporary base frame, the length and width of the required scaffold, and pack up level on bricks, blocks, etc. about 600 mm above the ground.
(2) Fix the four end standards to this frame with double couplers.
(3) Fasten ledgers to these standards and transoms to the ledgers where required, plumbing and levelling as work proceeds.
(4) Intermediate standards are now positioned, taking any sag out of the ledgers as necessary.
(5) Longitudinal and cross braces can now be fixed and the base frame removed.

Note Both types of scaffold must be securely tied to the building at least every 4 m vertically and 6 m horizontally. This is usually carried out in one of two ways

(1) with transoms passing through window or other openings, connected to tubes fastened with double couplers at right-angles to these inside the structure and close up against the wall (figure 6.22); or
(2) with short lengths of tube wedged in window openings with reveal pins; not more than 50 per cent of the ties may be of this type (figure 6.24).

Where openings are non-existent and the height of the scaffold is limited, the scaffold should be strutted from the ground with raking tubes inclined towards the building (figure 6.25).

A reliable method of securing a scaffold to a building where there are no openings for through ties or reveal ties is to use a ringbolt. The Hilti ringbolt, for example, has a shank length of 127 mm, a 16 mm thread and an internal ring diameter of 54 mm (see figure 6.26).

These are not suitable for fixing to the outer leaf of a cavity wall as this would result in the load from the scaffold being transferred to the wall ties, which are incapable of carrying such loads. The ringbolts should be attached to the structural frame of the building such as an edge beam or vertical column. A hole must be bored to the required depth, depending on the situation, and after cleaning out an anchor is inserted into which the ringbolt can be screwed hand-tight only (figure 6.27).

By providing ringbolts at the required spacings (refer to BS 5973) a length of scaffold tube is passed through, horizontally or vertically and the scaffolding can be securely fixed to these (see figure 6.28). When the scaffolding is taken down on completion of the works the ringbolts are unscrewed, the anchors left in place and the holes can be pointed up with matching mortar or plugged with the filler caps provided.

Figure 6.26

Figure 6.27

Figure 6.28

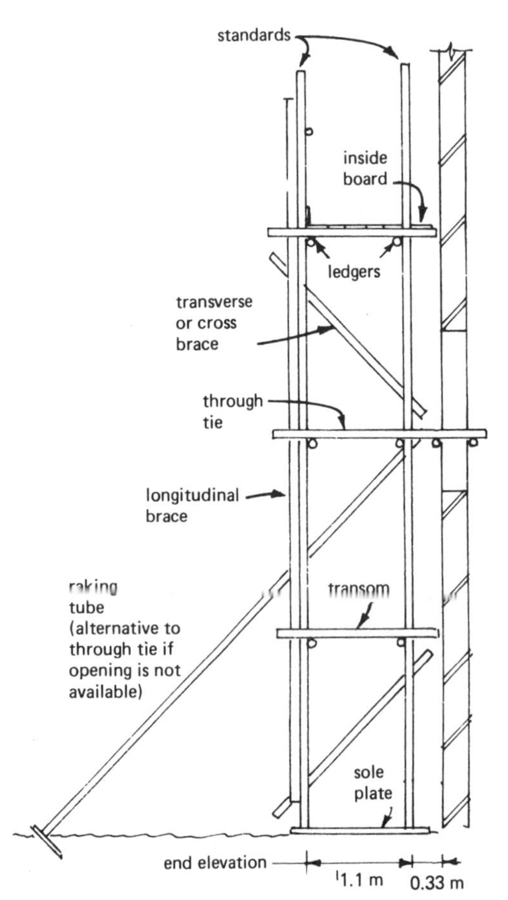

Figure 6.25 End elevation of independent scaffold

Widths of Working Platforms

Where an operative is liable to fall more than 2 m, the Construction (Working Places) Regulations lay down minimum platform widths of not less than

(1) 625 mm (loosely described as three boards wide) where the platform is used as a footing only, that is, not for depositing materials
(2) 850 mm (loosely described as four boards wide) where used for working from, and for depositing materials
(3) 1 m where used to support a higher platform

(4) 1.3 m where used for dressing stone
(5) 1.5 m where used to support a higher platform, and for dressing stone.

Responsibility for Scaffolds

Where a scaffold is erected by one employer and used by, or on behalf of, another employer, it is the responsibility of the first mentioned to ascertain that the scaffold and the materials from which it is constructed are sound and stable and that the Construction Regulations are kept to throughout.

Treatment and Storage of Equipment

Tubes should be stacked in racks in lengths, clear of the ground, and protected against the weather. Fittings should be cleaned and lightly oiled and stored in separate bins. When a scaffold is being dismantled, tubes should be carefully lowered to the ground since bent tubes may not be re-used and fittings require at least as much care to prevent loss or damage.

Trestle Scaffolds

Many types of adjustable steel trestle are available. Those shown in figure 6.29 are by S.G.B. Ltd and are designed in three widths to take three, four or five boards. The advantages of using trestle scaffolds include the following.

(1) They are light in weight, but strong.
(2) They are easily and quickly set up by one man.
(3) The widely splayed foot give stability, the legs being immovable when in use because of the fixed locating lugs.
(4) Adjustment of height is simple and positive and each of the four sizes available will extend to at least three-quarters of the initial height.
(5) The rests for the boards are flat and the pins are securely attached to the frames and cannot be lost or misplaced.
(6) Storage is facilitated by turning the splayed feet through 90°.
(7) They are very useful for single internal lifts in housing.

The Construction Regulations appertaining to trestle scaffolds lay down a maximum height of 4.5 m, which is generally considered to be three lifts; the trestles should be adequately braced to prevent sideways movement. Figure 6.30 shows a trestle scaffold with a raised platform for depositing materials, which eliminates a lot of the bending normally associated with bricklaying.

Figure 6.29 S.G.B. adjustable steel trestles

Figure 6.30

Tower Scaffolds

An independent tower scaffold, apart from the necessary ties, stands completely free from buildings and is mostly used for overhead maintenance work where only a small working area is required.

It consists basically of standards, ledgers, transoms, diagonal bracing and plan bracing. Some of the requirements for independent tower scaffolds are as follows.

(1) Standards should be on base plates with sole plates where required, and should be no more than 2.5 m apart.
(2) Ledgers must be fixed to standards with double couplers, the normal height of each lift being 2 m, which provides headroom for working on intermediate platforms where required.
(4) Transoms should be fixed to standards where possible with double couplers, intermediate transoms being fixed to ledgers with putlog couplers.
(5) Diagonal bracing should be fixed on all sides with plan bracing at the base and other levels where necessary for rigidity. Figure 6.31 shows the first lift of a tubular tower scaffold.
(6) Where the height of the tower is more than three-and-a-half times the shortest side it must be adequately tied.

Mobile Towers (figure 6.32)

Relevant details are similar to those for independent tower scaffolds, except for the base plates, which are replaced by lockable castors. Other requirements are as follows.

(1) The maximum height of internal towers is three-and-a-half times the shortest base dimension, while that for external towers is three times these dimensions.
(2) When the scaffold has to be moved, force should

Figure 6.31 First lift of an independent tower scaffold

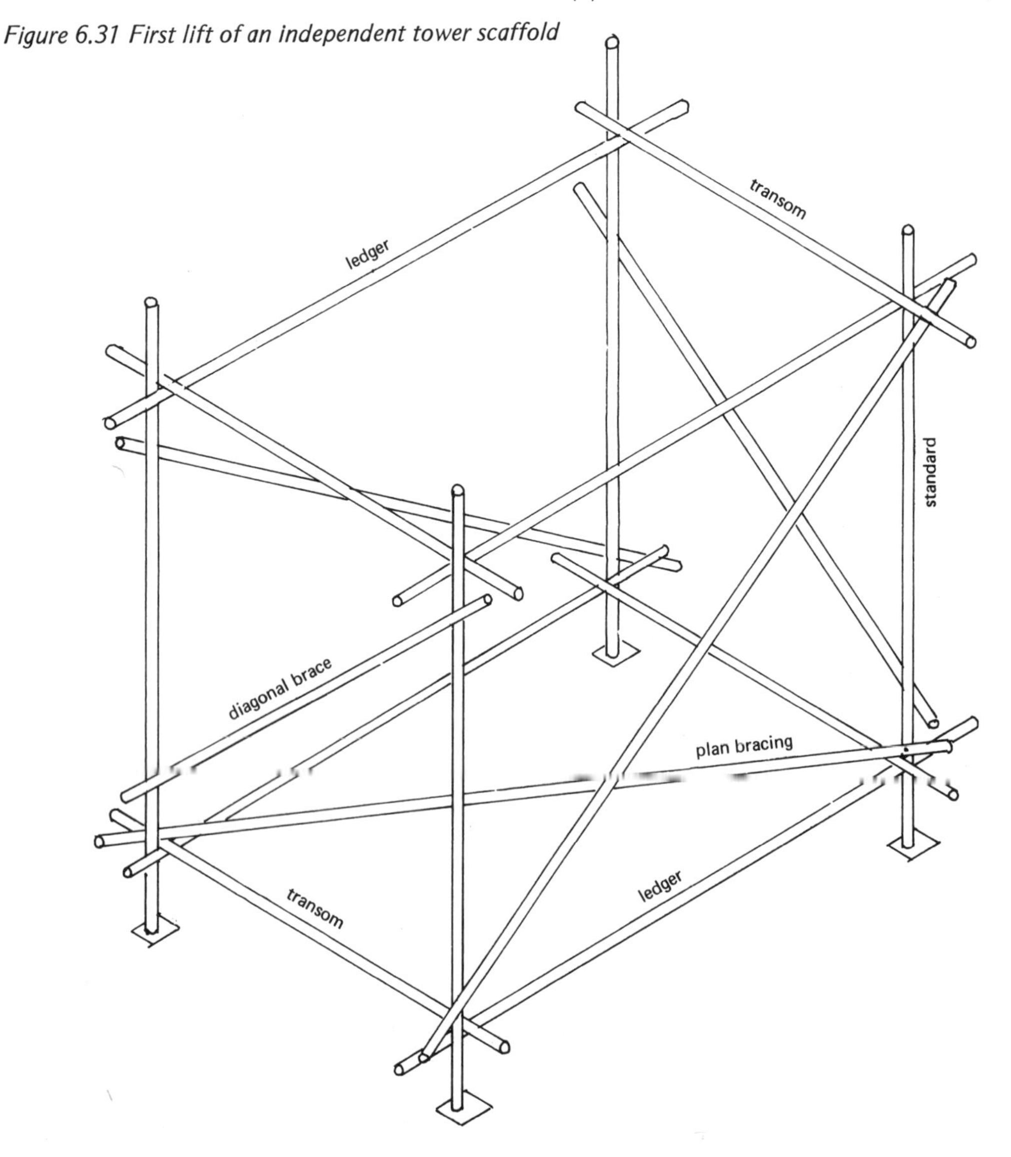

be applied near the base; do not pull or push it along while standing on a platform.

(3) Use only on ground that is firm and level.

(4) Brakes must be locked on when in use and the tower should be tied to the structure whenever possible.

(5) Only one working lift should be in use at a time.

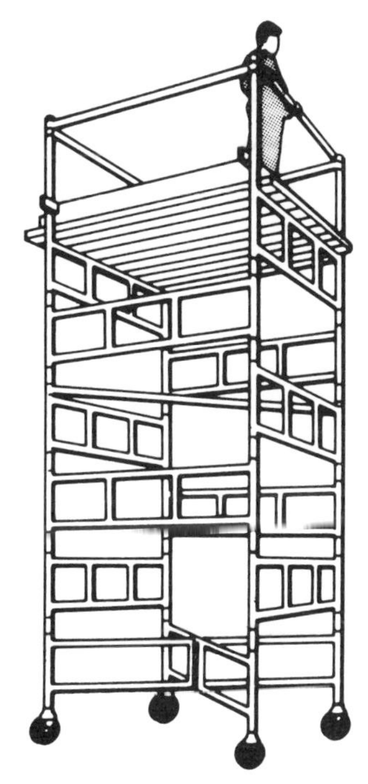

Figure 6.32 Lightweight access tower scaffold

Framed or System Scaffolds (figures 6.32 and 6.33)

These usually consist of metal H-frames constructed from patent welded units, which can be quickly interlinked to form independent or tower scaffolds. Each frame consists of two vertical members and one or two cross members. They are erected by joining two frames together with cross braces or ties and the height is increased by locating the next pair of frames over the spigots on the lower frames. See figure 6.33, which shows S.G.B. Sureframe scaffold, which, although a product of the 1960s, is still very popular, especially for support work. More recent developments in system scaffolding include, for example, Kwikstage and Cuplock, which are representative of the late 1970s.

With the Sureframe scaffold (figure 6.33) no couplers or fittings are necessary for the ledger and transom assembly, which carries the ends of abutting boards. The function of the ledgers is to span between frames to receive and support the transom unit.

LADDERS

Timber pole ladders, and timber or aluminium extending ladders are available for use as access to scaffolds, the former because of its strength being more common on building sites. The following is a brief summary of the Construction (Working Places) Regulations appertaining to the use of ladders.

(1) Ladders are to be sound, of adequate strength, properly maintained and to have neither missing nor defective rungs.

(2) Rungs are to be properly fixed to the stiles and must not rely on nails.

(3) Where possible ladders must be fixed at their upper end to the scaffold and must extend at least 1 m above the platform unless some other handhold is provided.

(4) where top fixing is impracticable ladders should be fixed near the bottom and must not be allowed to sway or sag unduly.

(5) Both stiles must be equally and firmly supported and must not be stood on loose bricks or other loose packings.

(6) Landings must be provided every 9 m in height.

(7) Ladders must not be painted to hide defects.

Ladders should be inclined at an angle of or near 75°, commonly referred to as four up, one out (figure 6.34). While a suitable procedure is to rear the ladder against the scaffold it is better to prevent the overhang at the top from encroaching on the scaffold by rearing the ladder against an extended putlog or transom, which must be secured to the ledger with a double coupler (figure 6.34*b*).

LIFTING EQUIPMENT

The Wedge

This is a simple inclined plane, usually of timber, which is moved forward by a series of hammer blows while the body (the object to be raised) remains – as it were – in a fixed position. Figure 6.35 shows the forces involved, and a comparison of figures 6.35 and 6.36 makes it obvious that, although the amount of lift is less, the narrower the wedge, the less effort is required. Wedges are often used in pairs (figure 6.37) and a series of blows to each wedge will raise the object without tilting occurring.

The Screw Jack

This too is an inclined plane in spiral form, in which the lever is rotated about its vertical axis in order to raise the load, which is placed on the swivel head. The

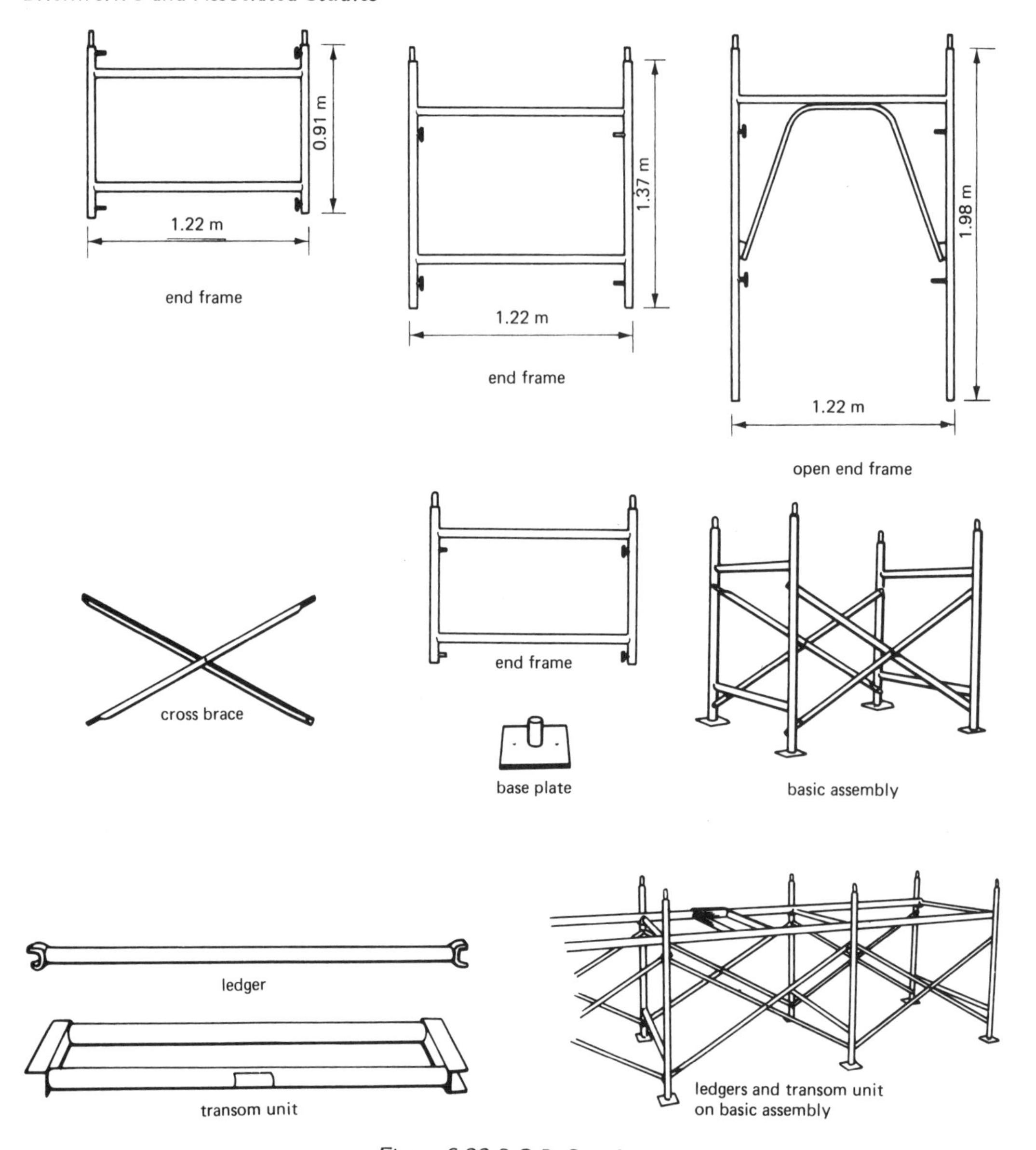

Figure 6.33 S.G.B. Sureframe

distance between the top of one thread and the top of the adjacent thread is known as the pitch of the screw. The length of the lever may be as long as 900 mm in some cases and it should be clear from figure 6.38 that one complete revolution of the lever will raise the swivel head a distance equal to the pitch.

With the screw jack little strength is required to lift heavy loads. An Acrow prop, for example, works on this principle, and although the lever is only about 225 mm long the lifting and supporting capacity is well known in the construction industry.

As is the usual case with lifting equipment, however, we are not getting something for nothing since the effort is applied at the end of the lever, and this has to be rotated through a large distance to raise the load fractionally. For example, if the pitch of the screw is 5 mm and the length of the lever is 900 mm, the end of the lever moves through a distance of more than 5½ m to raise the load 5 mm; and with an Acrow prop, although the lever is only 216 mm long, it turns through a distance of nearly 1.6 m to raise the load 6 mm (dimensions converted).

Simple Pulleys

With a pulley it is possible for a man to raise an object several times his own mass, on to a scaffold with the minimum of effort. The act of bending down and lift-

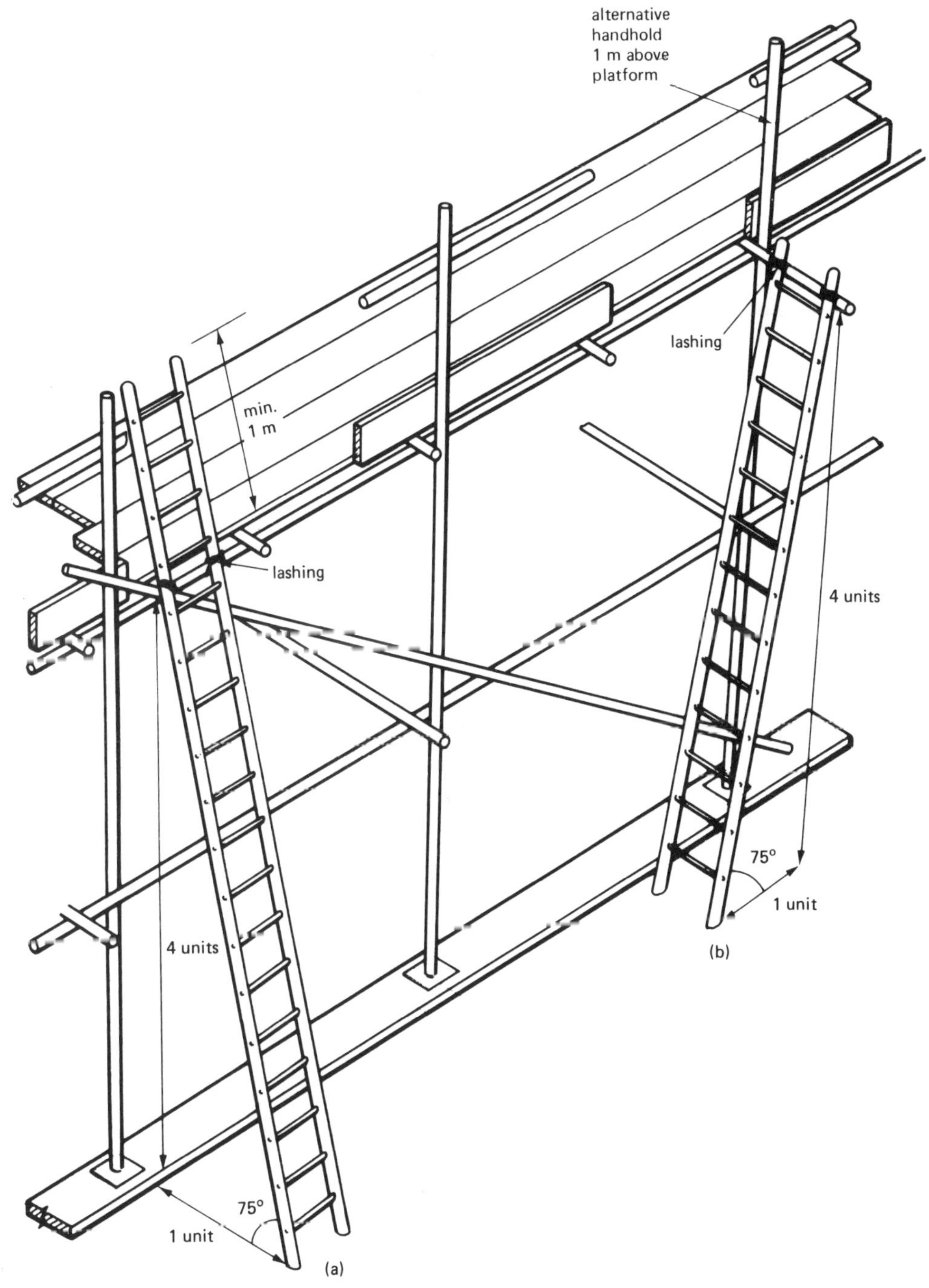

Figure 6.34 Isometric drawing of a putlog scaffold showing two methods of using ladders

ing a heavy object is difficult and may be dangerous, and it is much better to raise the object by heaving downwards on a rope. The simplest form of pulley is the gin wheel, which consists of a single wheel over which the rope is passed (figure 6.39*a*). The hook must be firmly secured to an extended putlog or transom with a figure-of-eight wire lashing turned at least five times round the hook and arranged so that the hook hangs 75 or 100 mm below the tube.

With an independent scaffold the support tube must be connected to both standards (figure 6.5), and with a putlog scaffold, the support tube should be connected to a standard and braced back to the level of mature brickwork. Support tubes to gin wheels

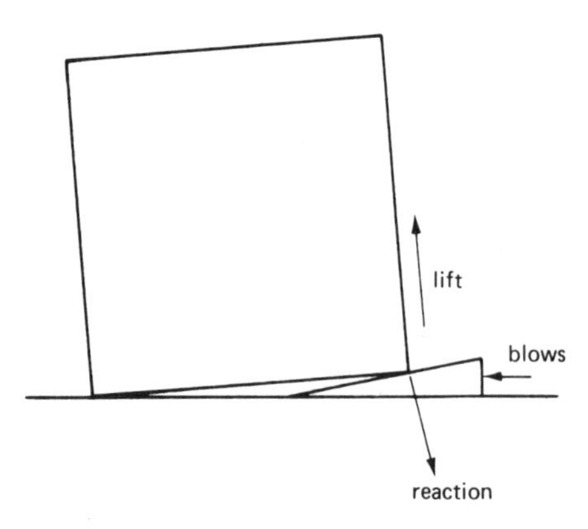

Figure 6.35 Use of narrow wedge

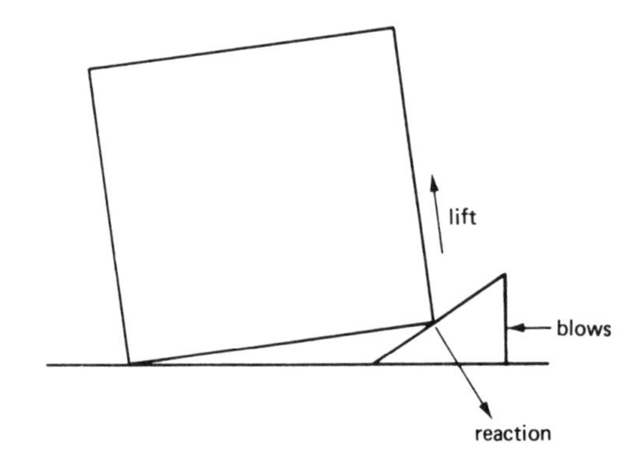

Figure 6.36 Use of steep wedge

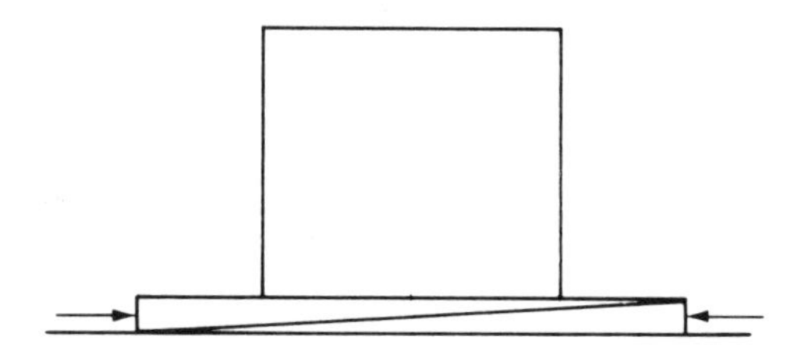

Figure 6.37 Use of folding wedges

should be placed as near to a positive tie as possible. This load is then attached to one end of the rope and the effort is applied at the other.

With a gin wheel a man cannot raise an object as heavy as himself and therefore the main reason for using this lifting appliance is that it is easier and safer to raise a heavy object by pulling downwards on a rope than it is to pull or carry it up on to a scaffold, provided that there is someone at the top to unload.

As already explained, to raise a mass of 50 kg, which creates a force of 500 N (50 x 10), the operative must be heavier than this; how much heavier depends to a large extent on the condition of the gin wheel

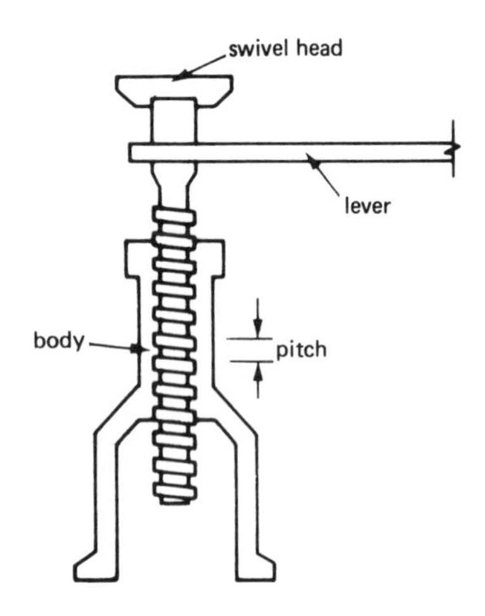

Figure 6.38 The screw

and the rope being used. Assume that 60kg is sufficient (600 N). The mechanical advantage (MA) of a pulley is found by dividing the mass to be raised by the effort required to raise it, and calculations should be carried out in newtons. Since the effort required is a force this will be given in newtons and if the mass is given in kilograms this must be multiplied by 10 (9.81 actually) to change it to newtons.

Example 6.1

If a mass of 50 kg can be raised by an effort of 600 N, what is the mechanical advantage?

$$MA = \frac{\text{mass} \times 10}{\text{effort}} \quad \begin{matrix}\text{(to change kg to N)} \\ \text{(already in N)}\end{matrix}$$

$$= \frac{50 \times 10}{600}$$

$$= \frac{500}{600}$$

$$= 0.833$$

On the other hand the mass may be given as a load in newtons in which case multiplication by 10 is unnecessary.

Example 6.2

If a load of 800 N is raised by a force of 900 N, calculate the mechanical advantage

$$MA = \frac{\text{force}}{\text{effort}}$$

$$= \frac{800}{900}$$

$$= 0.88$$

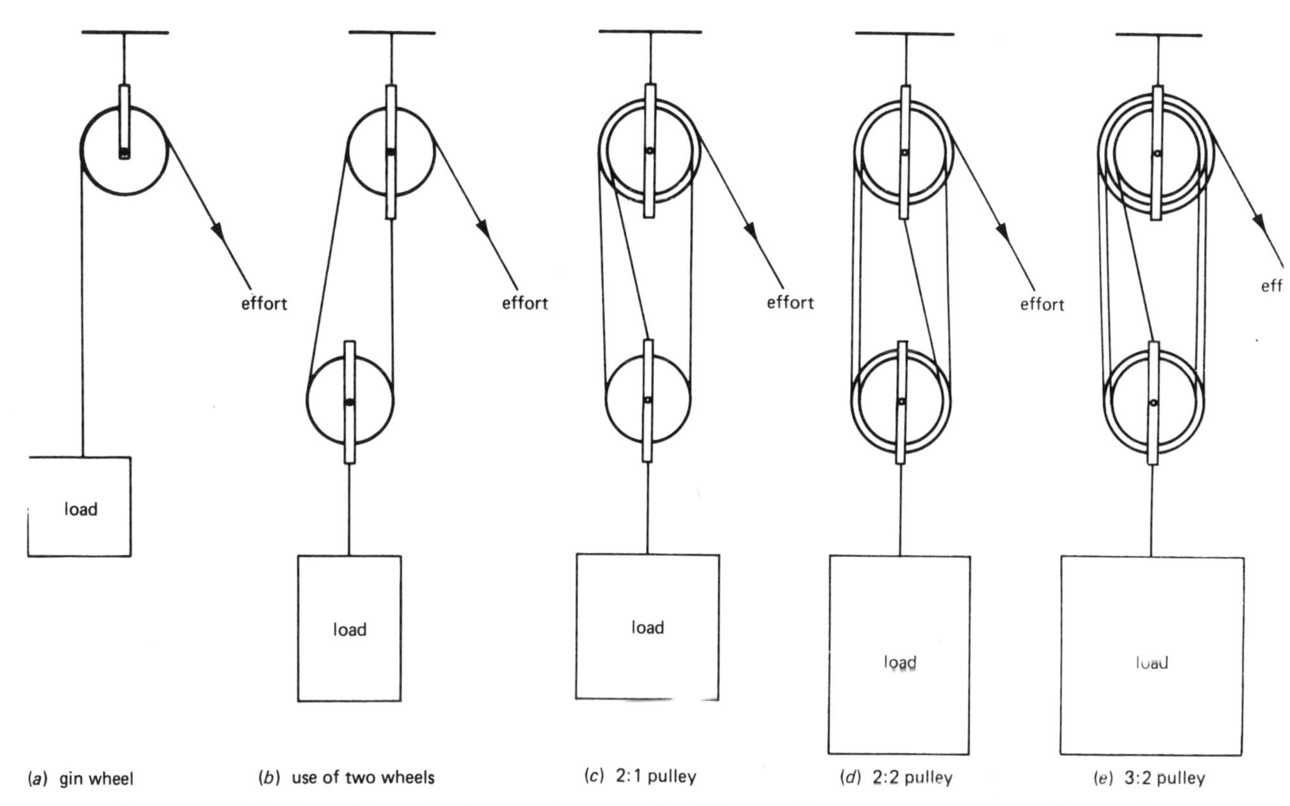

Figure 6.39 Pulleys (the wheels are shown with different diameters so as not to obscure the ropes)

The velocity ratio (VR) of a pulley is defined as the downward distance moved by the rope due to the effort, compared with the upward movement of the load. It will be obvious that with a gin wheel the amounts of upward and downward movement are equal and, therefore, the VR of a gin wheel is unity.

Note The VR of a simple pulley can also be found by counting the number of wheels used in the system. The efficiency of a pulley is calculated by dividing the mechanical advantage by the velocity ratio and multiplying the result by 100. In the case of example 6.1

$$\text{efficiency} = \frac{\text{MA}}{\text{VR}} \times 100\%$$

$$= \frac{0.833}{1} \times 100$$

$$= 83.3\%$$

Using More Wheels

When a hanging pulley attached to a load is supported by two ropes (figure 6.39*b*) the force in each rope is equal to half the force exerted by the mass. Therefore, to balance a mass of 100 kg an effort of 500 N (50 kg x 10) would be required. The VR would be 2 since pulling 300 mm downwards would only raise the load 150 mm (also, two wheels are being used). In figure 6.39*c* there are three ropes supporting the load and, therefore, if the load to be raised had a mass of 150 kg the force in each rope would be 500 N (50 kg x 10) and thus an effort of 500 N would balance a load of 150 kg, which creates a force of 1500 N (150 kg x 10). To raise the mass a little extra effort will be required.

Example 6.3

Assume that the mass to be raised is 180 kg and the force required to raise this is 800 N (80 kg x 10). Calculate MA, VR and efficiency.

$$\text{MA} = \frac{\text{mass} \times 10}{\text{effort}}$$

$$= \frac{180 \times 10}{800}$$

$$= \frac{1800}{800}$$

$$= 2.25$$

(This means that a man can raise 2¼ times his own mass when using this pulley.)

$$VR = 3 \text{ (number of wheels in system)}$$

$$\text{efficiency} = \frac{MA}{VR} \times 100\%$$

$$= \frac{2.25}{3} \times 100\%$$

$$= \frac{225}{3}$$

$$= 75\%$$

Consider figures 6.39*d* and *e*. While these are a little more complicated, still larger mechanical advantages can be gained with their use. In figure 6.39*e*, for example, a load having a mass of 250 kg can be balanced by a force of 500 N (50 kg x 10). This particular pulley has a velocity ratio of 5 (number of wheels used in the system).

The Scaffold Crane

This is loosely described as a powered gin wheel and it will raise a mass of up to 250 kg on to a scaffold platform. The crane is situated on and operated from the platform and the fixed slewing jib which is extended outwards must be adequately braced back to a standard before lifting starts. The load can only be moved laterally in a circle of fixed radius.

Figure 6.40

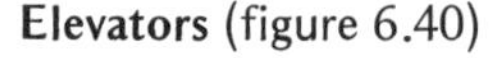

Elevators (figure 6.40)

Manual and hydraulic elevators are available that are either towable as a two-wheel trailer behind a car or lorry or can be mounted on a lorry chassis. They are very useful for building contractors, scaffolders, glaziers, slaters and tilers, etc. since suitably sized platforms are available for attaching to the elevator to carry anything from bricks to 6 m scaffold tubes. Figure 6.41 shows the top of an elevator transporting bricks.

The heights and angles to which elevators can be used vary greatly: those produced by Walter Somers, for example, vary in height from 8 to 30 m and in certain circumstances extensions to 40 m are possible. They can be erected by one man, and are ready for work to start in 10 minutes; with a man to load at the bottom and another to unload at the top, materials are very quickly transferred. Where materials are intended to be tipped at roof or other level this facility can be incorporated into the elevator (figure 6.42).

Figure 6.41

Figure 6.42 Combined tipping and rubbish bucket for tipping below and above

Hoists

A hoist consists of a horizontal platform which moves up and down vertical guides by a powered winch. The guides are normally tied back to the structure or scaffolding to provide stability.

The Construction (Lifting Operations) Regulations 1961, part V cover the requirements for hoists and these are summarised as follows.

(1) The hoistway must be protected by a substantial enclosure at all points where access is provided or where persons may be struck by moving parts.
(2) Safety gates must be provided at each landing for loading and unloading.
(3) An automatic brake is to be provided to support the platform in the event of failure of ropes or any other part.
(4) A safety device must be installed to ensure that the hoist cannot overrun its highest point.
(6) The safe working load is to be clearly marked and this must not be exceeded except for testing. A hoist to carry persons is to state the maximum number to be carried.
(6) A notice is to be placed on the platform stating that persons are not to be carried unless the hoist is so designed (part VI 48).
(7) The hoist is to be operated from one position (not from within unless part VI 48 is complied with) and if the operator cannot clearly see the platform arrangements for suitable operating signals must be made.
(8) No person under the age of 18 to operate a hoist, or give signals to an operator unless under direct supervision for the purpose of training.

Mobile Hoists

With this type of hoist the mast can be quickly lowered by two or three men and moved on two pneumatic tyres. It was developed mainly for housing and models range from those that can lift 250 kg to a height of 4 m, to the largest, which can raise 1 tonne to a height of 5 m. The mast is extendable in some cases, but over 6 to 8 m – according to the model – the mast must be tied to the building, in which case it can no longer be classified as mobile. Mobile hoists are very useful for serving two or three houses in close proximity since in this type of work the rate at which materials are used is usually relatively slow.

Fixed Hoists

As the name implies, this type of hoist remains in a fixed position throughout a contract and sizes vary from ½ to 3 tonnes capacity, which can be raised to heights of about 150 m. The platforms of these hoists can be side-slung as in the previous case, or centre-slung, in which case a tubular scaffold tower is erected to enclose it. To stabilise a fixed hoist it must be tied at regular intervals to the building.

The Lever

While on the subject of lifting appliances, it is perhaps appropriate to consider the use of levers at this point. These are extremely useful in certain situations – for example, where a heavy object has to be raised from the ground to place rollers underneath in order to move it to another place. There are in fact three different methods of using a lever, and these are known as the three orders of levers.

The first order of levers Consider figure 6.43. Here the fulcrum is placed between the object to be raised and the downward force required to raise it. This is the most common way of using a lever. It will be obvious from figure 6.43 that the closer the object is

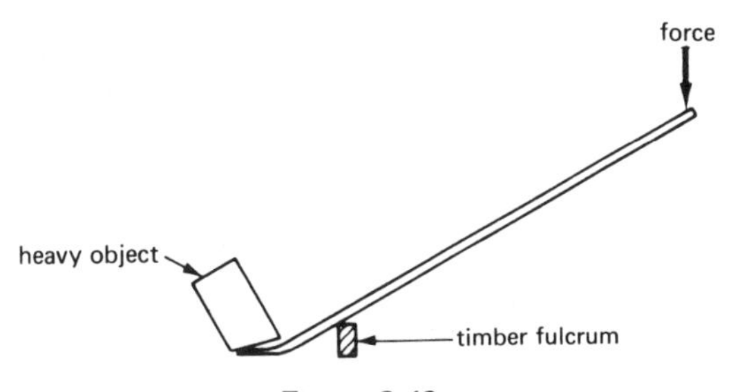

Figure 6.43

to the fulcrum and the longer the lever, the less effort will be required to raise the object. The object is raised because the clockwise moment about the fulcrum is greater than the anticlockwise moment.

Figure 6.44 shows the apparatus necessary for a simple experiment to study the first order of levers.

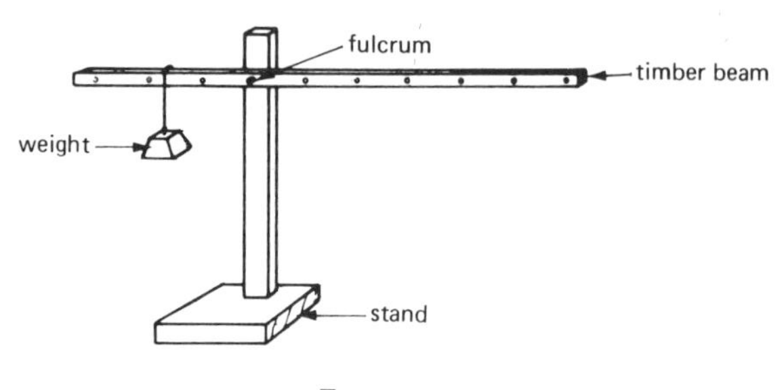

Figure 6.44

Experiment

Requirements for the experiment are a timber beam, measuring 38 x 25 mm, 1 m in length, with holes every 100 mm; one winged nut and bolt, a stand and various weights.

Method Set up the apparatus as shown in figure 1.40, fixing the fulcrum in a hole between positions 3 and 8 inclusive. Place a small weight on the short length of the beam to balance the self weight of the beam.

Start by hanging a 2 kg weight (creating a force of 20 N, assuming gravity is 10 m/s^2) at a distance of 600 mm to the right of the fulcrum, and a 4 kg weight (force 40 N), 300 mm to the left of the fulcrum. The beam will be seen to balance. This is because the clockwise moment (CM) is equal to the anticlockwise moment (ACM). That is

20 N x 600 mm = 1200 N mm (CM)

and 40 N x 300 mm = 1200 N mm (ACM)

Now place a 3 kg weight (30 N) at 500 mm from to the right of the fulcrum. What force is needed at 300 mm to the left for the beam to be in equilibrium?

30 N x 500 mm = 1500 N mm (CM)

and *x* N x 300 mm = 1500 N mm (ACM)

The force required is 50 N (5 kg).

A table can now be drawn up as shown in table 6.1 and different forces applied at varying positions on the beam.

Thus it can be stated that a moment is the turning effect of a force about a point and its value is determined by multiplying the size of the force by its distance from the point.

Forces are given in newtons (N) or kilonewtons (kN) and distances in millimetres or metres.

Figures 6.45 and 6.46 show two identical cantilevered timber beams, each supporting identical forces of 2 kN. It is obvious that if either beam were to fail under load, it would be the one in figure 6.46, since the force is acting at a greater distance from the beam support. The moment produced in figure 6.45 is

force x distance = 2 kN x 1 m

= 2 kN m

The moment produced in figure 1.42 is

force x distance = 2 kN x 5 m

= 10 kN m

Figure 6.47 shows a heavy plank supported by a trestle. A bag of cement is placed at the position shown. Would the plank remain in its position or would the force produced by the bag of cement force the right-hand side down?

This would be difficult to ascertain from the figure shown, but if the cement were moved to the extremity of the plank, that is, on the right-hand side, it would almost certainly over-balance. The weight of the cement has not changed but a change has occurred in its distance from the fulcrum. The moments produced by the bag of cement before and after moving are

position 1

moment = force x distance

= 500 N x 1 m

= 500 N m

position 2

moment = force x distance

= 500 N x 2 m

= 1000 N m

(assuming g = 10 m/s^2)

Figure 6.48 represents a beam resting on a fulcrum and carrying different forces at different positions from the fulcrum. Ignoring the self weight of the beam, will this arrangement produce a balance or will one end be lowered to the ground?

This can be determined by experiment or calculation. The calculation method would require the calculation of moments produced on each side of the fulcrum. Figures 6.49 and 6.50 show diagrams of the anticlockwise and clockwise moments respectively

Table 6.1

Left-hand side, ACM				Right-hand side, CM			
Mass (kg)	Force (N) (mass x g)	Distance from fulcrum (mm)	Moment (N/mm)	Mass (kg)	Force (N) (mass x g)	Distance from fulcrum (mm)	Moment (N/mm)
2	20	600	12000	4	40	300	12000
3	30	500	15000	5	50	300	15000

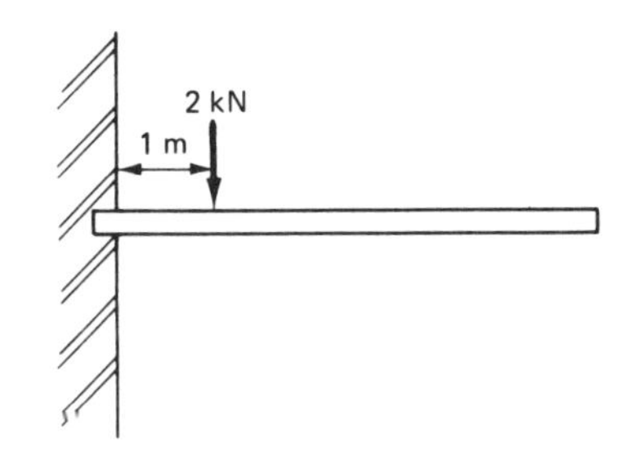

Figure 6.45

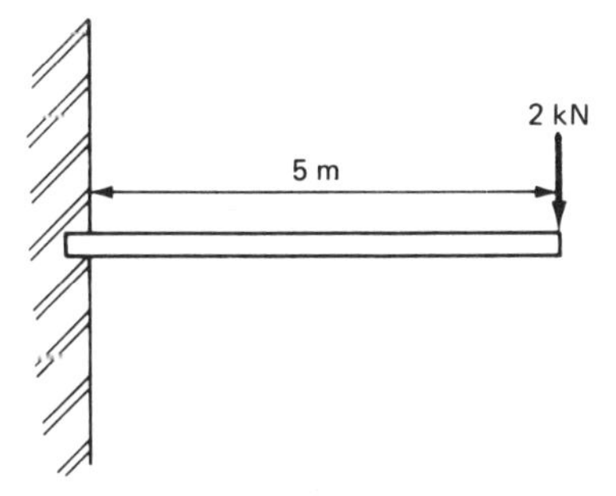

Figure 6.46

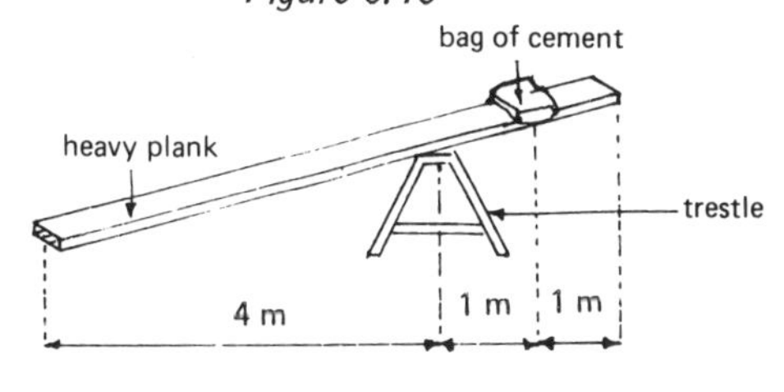

Figure 6.47

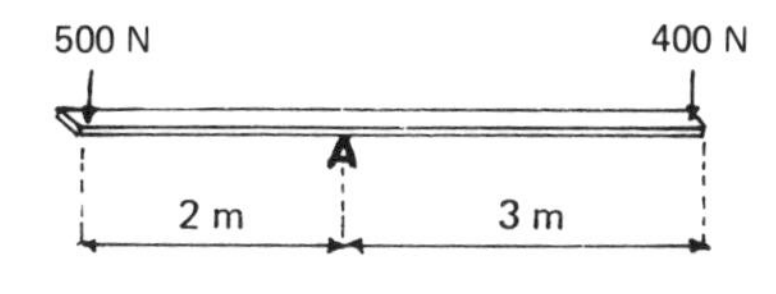

Figure 6.48

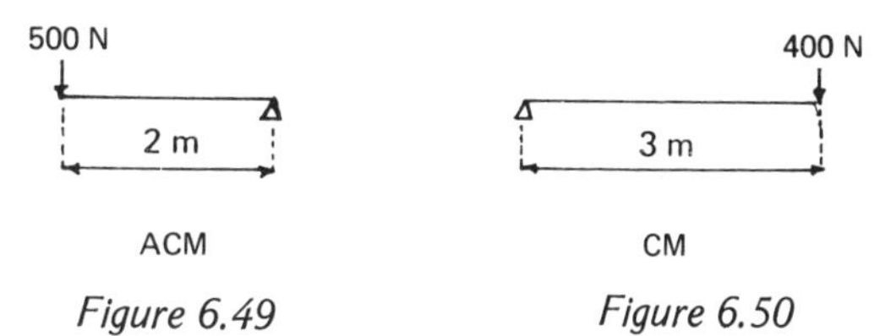

Figure 6.49 *Figure 6.50*

Anticlockwise moment = force x distance
= 500 x 2
= 1000 N m

Clockwise moment = force x distance
= 400 x 3
= 1200 N m

This result would indicate that the right-hand end of the beam would be lowered and resting on the ground.

First Order of Levers

Example 6.4

Calculate the effort required to raise the load shown in figure 6.51.

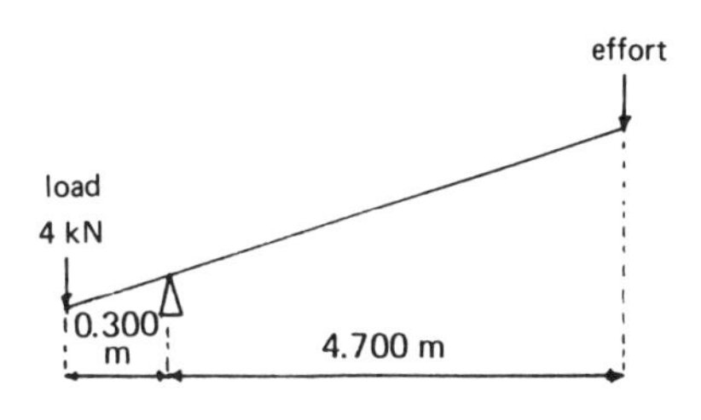

Figure 6.51

Clockwise moments (CM) = anticlockwise moments (ACM)

Therefore

$$\text{effort} \times 4.7 = 4 \times 0.3$$
$$= 1.2$$

Therefore

$$\text{effort} = \frac{1.2}{4.7} \quad \text{(divide both sides by the coefficient of 'effort')}$$
$$= 0.26 \text{ kN}$$

Thus an effort of 0.26 kN (260 N) will balance the load and any addition to this will cause the load to rise.

Example 6.5

Calculate the effort required to raise the load shown in figure 6.52

$$\text{CM} = \text{ACM}$$

Therefore

$$\text{effort} \times 1.4 = 7 \times 0.6$$

Therefore

$$\text{effort} \times 1.4 = 4.2$$

Therefore

$$\text{effort} = \frac{4.2}{1.4}$$
$$= 3 \text{ kN}$$

Therefore any effort over 3 kN will cause the load to rise.

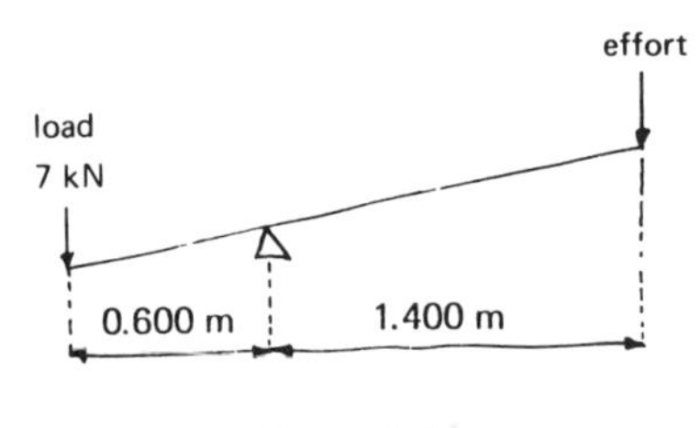

Figure 6.52

Second Order of Levers

A method of raising a heavy object that is already clear of the ground is to place a plank or crowbar under the object with one end resting on a firm surface and lift it upwards to form contact with the object. An upward force is then applied at the other end of the lever (figure 6.53).

The wheelbarrow is an excellent example of the use of the second order of levers (figure 6.54).

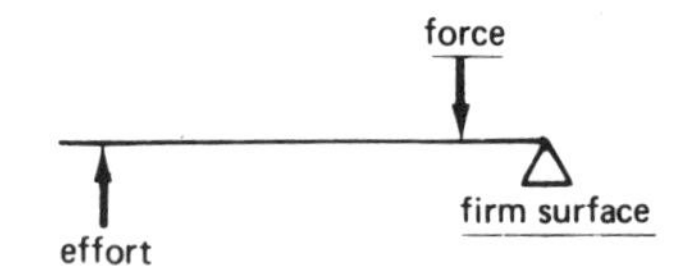

Figure 6.53

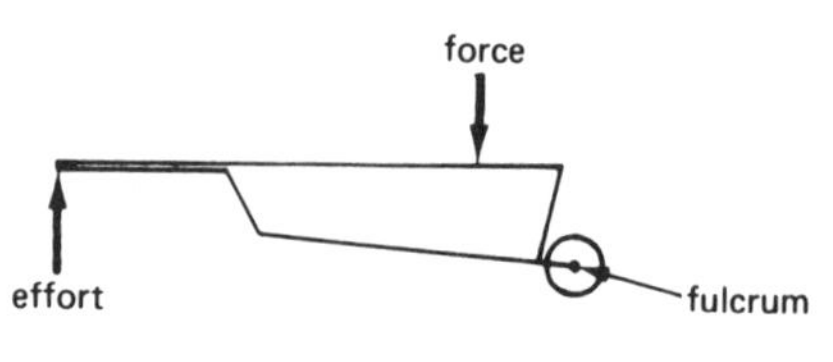

Figure 6.54

Example 6.6

Calculate the effort required to raise the load shown in figure 6.55.

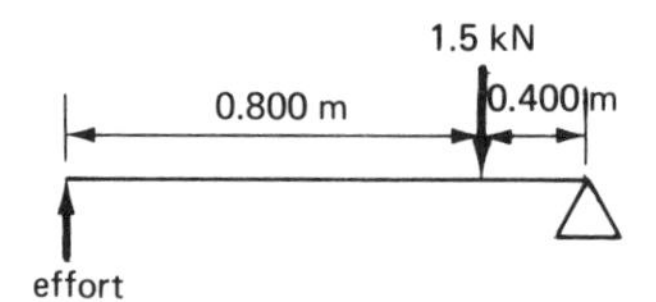

Figure 6.55

Note Moments are always calculated from the fulcrum.

Clockwise moments = anticlockwise moments

Therefore

$$\text{effort} \times 1.2 = 1.5 \times 0.4$$
$$= 0.6$$

Therefore

$$\text{effort} = \frac{0.6}{1.2}$$
$$= 0.5 \text{ kN}$$

Any effort greater than 0.5 kN will raise the load.

Third Order of Levers

Figure 6.56 shows the arrangement for the third order of levers. In this example the effort required to raise the load is always greater than the load itself.

Figure 6.57 shows this to be a very convenient arrangement.

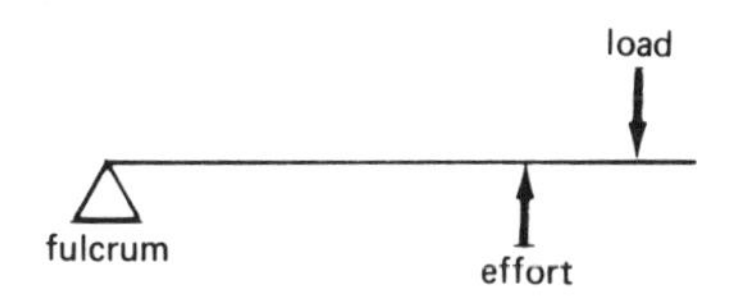

Figure 6.56

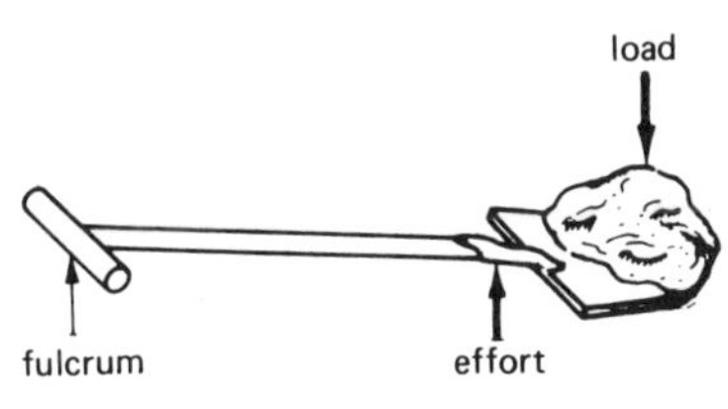

Figure 6.57

Example 6.7

Calculate the effort to balance the load shown in figure 6.58.

Clockwise moments = anticlockwise moments

Therefore $50 \times 1000 = \text{effort} \times 800$

Therefore

$$50\,000 = \text{effort} \times 800$$

Therefore

$$\text{effort} = \frac{50\,000}{800}$$

$$= 62.5 \text{ N}$$

It will be noted that the effort is in excess of the load.

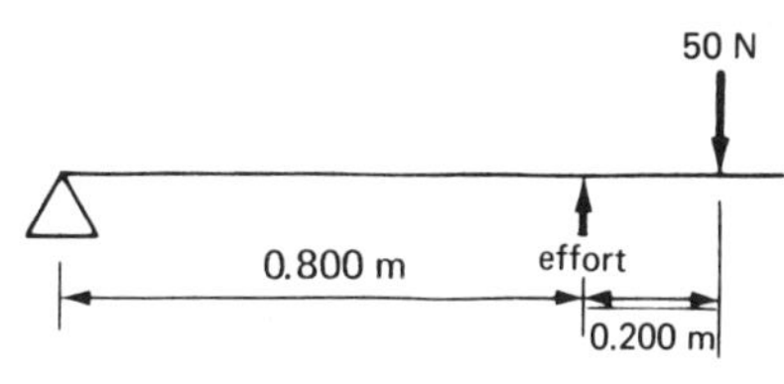

Figure 6.58

7
MAINTENANCE, REPAIR AND FIXING EQUIPMENT

The construction industry provides its craftsmen with a greater variety of work than any other industry. In the construction of new buildings, which is classified as new work, employment is provided for approximately 25–35 per cent of the labour force. In maintenance, repair and alteration, work is constantly provided for the majority of building trade workers.

In the work of maintenance and repair we find that situations, buildings and materials are never identical, and techniques that are suitable for a particular situation may not be universally applicable because so many other factors have to be considered. The bricklayer craftsman should, therefore, be equipped with the skills that will enable him to cope with the many different problems and varying situations in which he may become involved.

REPLACING DEFECTIVE FIREBACKS

All types of fireback may become defective because of the following causes

(1) abrasion and resistance to heat, which is termed normal wear and tear
(2) movement of the surrounding infill material, the structural hearth, tiled hearth and tiled surround: if any movement caused is not accommodated it will result in damage to the fireback.

It is recognised that sectional firebacks are capable of resisting movement better than the single unit or one-piece fireback, and obviously the four-section fireback is more capable than one comprising two sections.

When the craftsman is required to renew an existing defective fireback, it is important to inspect and determine the area of damage and the cause of the trouble.

Figure 7.1*a* shows areas of disintegration in the fireback, with spalling, laminating and sometimes fractures originating from the worn areas. Where only slight cracking and a minimum amount of spalling is found, repair can be effected by pointing with patent fire-cement, otherwise complete reinstatement is required.

Figure 7.1*b* shows a very large crack occurring at the centre of the fireback, running from top to bottom. This is caused by lateral pressure on the fireback, resulting from the absence of a movement joint around the back of the fireback, with consequent pressure from the infill material.

Figure 7.1*c* shows a large horizontal crack occurring about half way up the fireback. This is caused by uneven pressure on the fireback, and is due to movement of the infill material and to the absence of a movement joint between the fireback and the tiled surround.

Figure 7.1*d* shows a deep vertical centre fracture and also fractures in the side cheeks. The causes of this defect are the total absence of any movement joints, badly placed infill material and sometimes the wrong type of infill material.

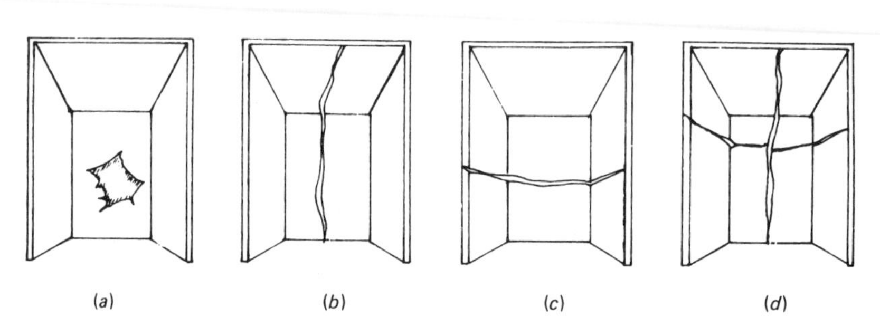

Figure 7.1 Defective firebacks

Removing the Fireback

Where the fireback consists of a one-piece unit, removal can easily be effected with the lump hammer and an 18 mm-diameter cold-steel chisel at least 250–300 mm long. The cutting-out operation should be started around the central fracture and the side cheeks can be taken out after the centre of the back has been removed.

It should be noted that before any of the above operations are begun, it is essential to provide complete protection for the tiled surround and hearth. This can be done with sacking over the hearth and drapes over the tiled surround, which will prevent any chipping or spalling of the tiles should they be struck by particles from the fireback.

Removing the Infill Material

Removing the infill material may require the craftsman to use a lightweight percussion drill with a chisel end, or the lump hammer and a 25 mm-diameter cold-steel chisel at least 450 mm long.

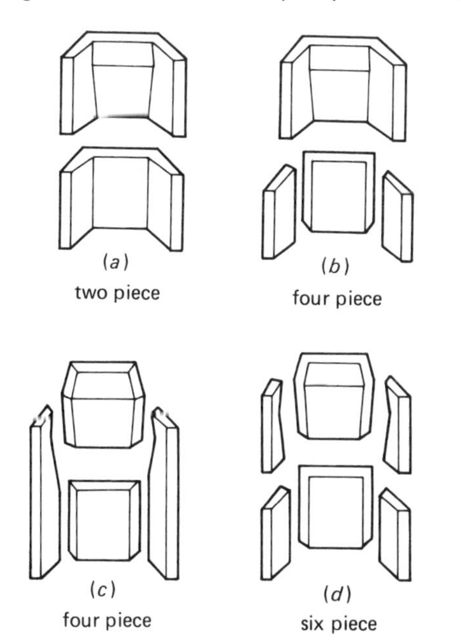

Figure 7.2 Firebacks complying with BS 1251: Part 1

Fixing the New Fireback

A four- or six-piece fireback is to be preferred in this situation. While it may be just possible to use a one- or two-piece, this is difficult, although not impossible, if the fireplace recess has been constructed to the minimum dimensions (Volume 2, chapter 11, figure 11.6).

Assuming the four-piece has been selected, before it is placed in position expansion joints must be formed on either side at the back of the new tiled surround where the new fireback abuts. These joints consist of lengths of fibre glass rope which have been cut to the height of the new fireback and soaked in waterglass.

The two firecheeks and base back section (figure 7.2*c*) are stood in position with fibre glass string between each joint, and a further expansion joint is formed by wrapping corrugated cardboard around the back. This material will later disintegrate with the heat, leaving an open joint between the infill and the fireback which will allow movement to take place without causing stresses to be set up and eventual cracking to occur.

The infill should consist of broken brick and lime mortar, or a weak concrete mix such as 1:4:10 (figure 7.3). If bricks and mortar are used, they should be built up as high as possible all round before bedding the top back section — again on fibre glass string. It is important that the infill is not tight up against the fireback. The throat can now be formed, the chamfered lintel bedded in position and made good to the gatherings.

If the fireplace recess is higher than usual, as in older properties, a hole will need to be knocked through above the tiled surround to complete the bricking up. This must be made good later and replastered (figure 7.4).

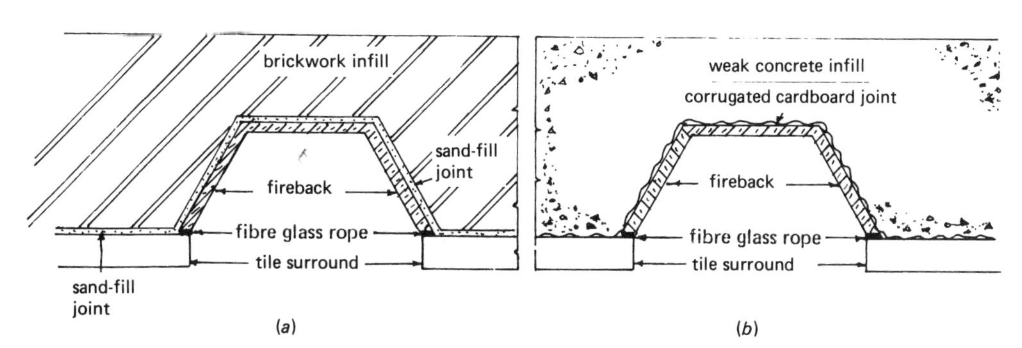

Figure 7.3 Plan of firebacks showing infill and movement joints

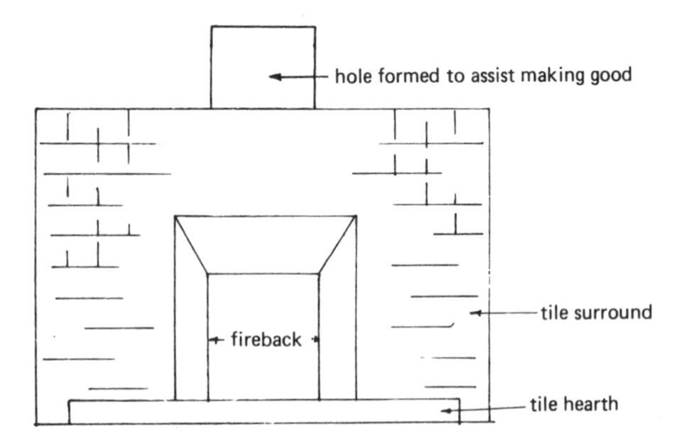

Figure 7.4 Method of making good the infill behind and above fireback

The joints between the fireback sections should be pointed up with fire cement.

DEFECTIVE CHIMNEY STACKS

Work Before Inspection

A suitable type of ladder should be erected and placed in position to provide easy access to the roof. Because materials may be taken on and off the roof, it is advisable to use a pole-sided type, which should be secured at the required height above the eaves and at the correct angle (figure 7.5). Crawling boards or crawling ladders are then placed in position on the roof to allow inspection of the chimney stack.

Inspection

This can now start with the chimney capping and then the entire brickwork of the stack. The ability to recognise and determine the following factors is a prerequisite for inspection.

(1) the condition of the capping and the capping material
(2) the type and condition of the bricks and mortar used for the stack
(3) any visible fractures in the brickwork or flaunching
(4) whether the damp-proof course is effective, whether there are flashings and soakers, and in what condition.

Faults and Failures

The following faults and failures are recognised as being common and are often found on existing houses and old buildings

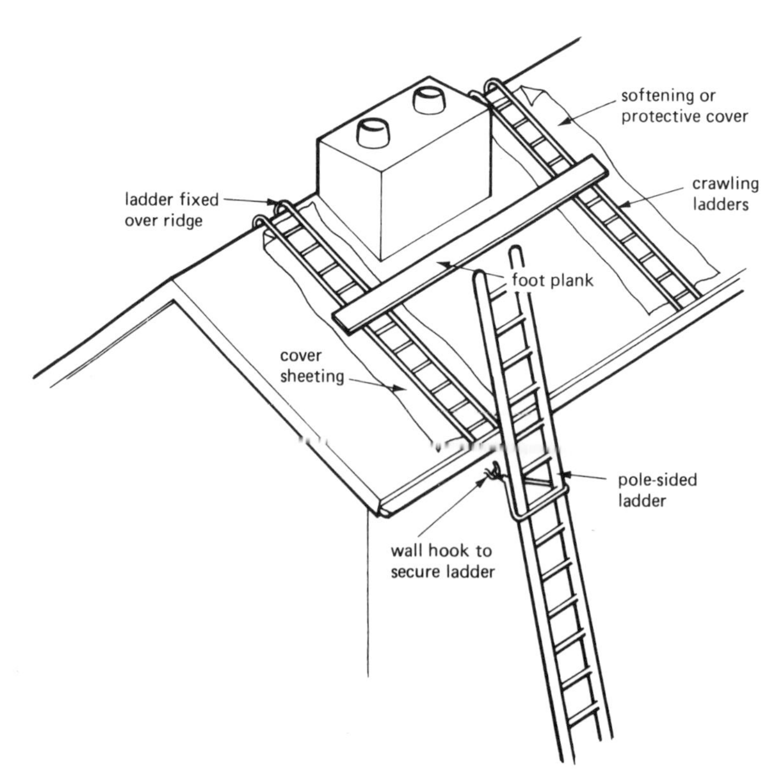

Figure 7.5 Requirements for inspection of chimney

(1) leaning chimney pots
(2) loose capping brickwork
(3) cracked and loose flaunching
(4) fractured stacks, showing visible cracks
(5) spalling and lamination of the stack bricks
(6) deterioration of the mortar joints
(7) leaning chimney stacks
(8) ineffective d.p.c.

The first seven failures are often caused by sulphate attack and may be accelerated by the ineffectiveness of the chimney capping where it has insufficient projection or is badly weathered and formed with weak materials. Fault 8 can be considered to be a contributing factor to some of the other failures.

Conclusion of the Inspection

Where the condition of the stack is poor, and flue liners are obviously not present, the entire stack should be taken down to three to four courses below roof level and rebuilt with suitable new material, with flue liners inserted during construction.

Protection

Before any work is carried out on the chimney stack, the fireplaces below should be inspected and sacking should be inserted at the fireplace throat to prevent debris entering the room. All roof work should be covered and protected in the area of the stack; precautions should also be taken to protect any work, or people below roof level.

Equipment

Where the chimney is of considerable height, scaffolding must be erected around the stack. Timber is normally used for roof work, although other methods can be used to provide the necessary working platform (figures 7.6–7.8).

Taking Down for the Stack

This should be done with considerable care, the dismantling starting with the chimney pot and flaunching, then the stack itself. All materials should be removed and taken down to ground level, then placed in a position where they will not impede any building operations and also where they can easily be removed from site.

Rebuilding the Stack

Materials

The type of bricks and mortar should be determined with considerable care, taking into account the previous failures, the necessity to combat sulphate attack, prevent weather penetration and blend in with the appearance of the building. Obviously, flue liners

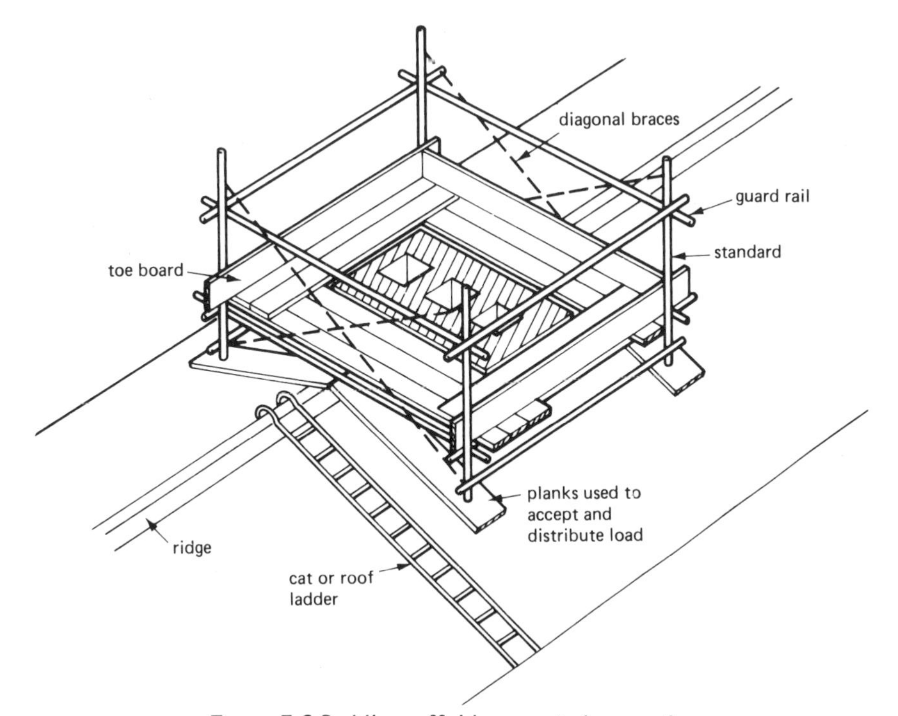

Figure 7.6 Saddle scaffold supported on roof

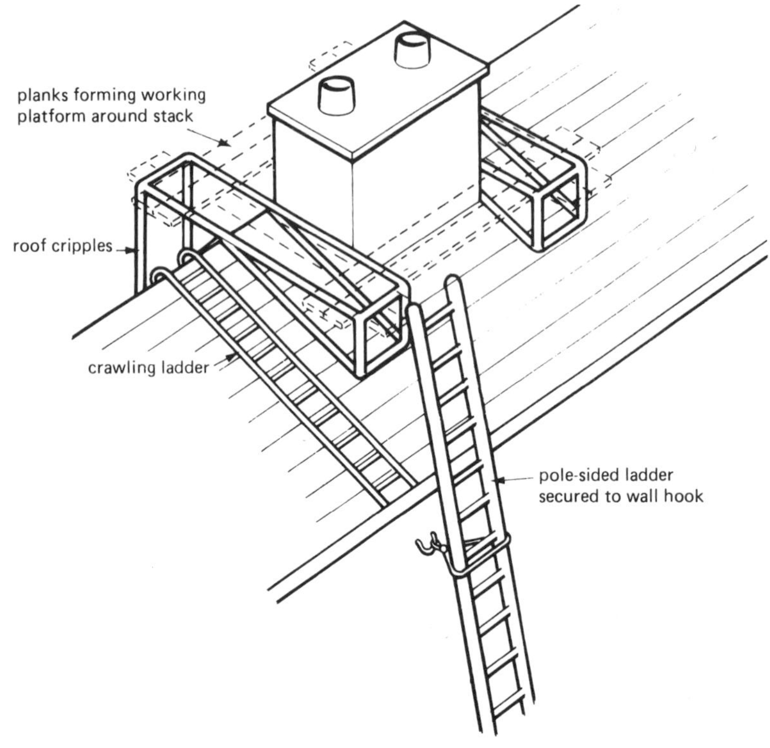

Figure 7.7 Working platform around stack formed with roof cripples (for inspecting only)

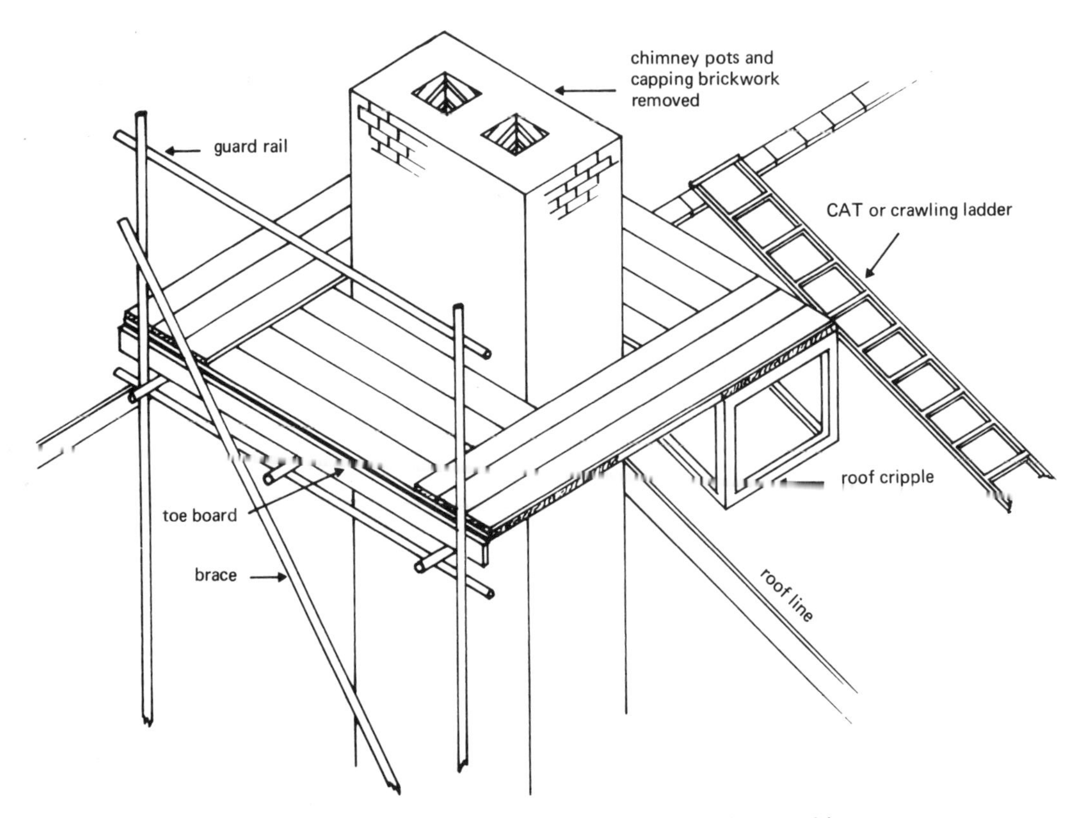

Figure 7.8 Working platform around external stacks on gables

must be inserted at the position where rebuilding is begun.

Procedure

At the required position below roof level, rebuilding is begun. When the work reaches 150 mm above the lowest point of intersection between roof and stack, a tray d.p.c. should be inserted; the mortar joints around the base of the stack above roof level should be raked out to accommodate the flashings. Flue liners must be inserted up to chimney pot level and these must be bedded and jointed in the same mortar that is being used for the brickwork of the stack. The surrounding space within the stack should be filled with solid material.

At capping level, adequate protection should be provided, oversailing courses should be formed or a precast concrete slab used to form the entire capping as one unit. The chimney pot can then be inserted and fixed. The terminal should be selected with the following in mind:

(1) the pot should match the liners used – that is, both circular or both square on plan
(2) it should complement the chimney stack and building below.

After rebuilding is completed, the sacking can be removed and the flue examined for draught by lighting a low fire. If the results are good, the scaffolding above roof level can be taken down, and the roof completely cleaned off, with all protecting covering removed, and the building waste taken away.

DEFECTIVE BRICK-ON-EDGE WINDOW SILL

It should be recognised that window sills that are constructed with bricks and mortar can become defective as a result of any of the following

(1) poorly selected bricks
(2) weak or poorly mixed mortar
(3) large mortar joints
(4) poor construction due to bad workmanship
(5) water remaining on top of the sill
(6) insufficient projection.

To construct a good weather-resistant brick sill, it is important that the following considerations are complied with.

(1) The bricks used for the sill should enhance and complement the building, but should be as dense and non-absorbent as possible.
(2) The mortar used to bed and form the sill should be of the same density as the bricks.
(3) Mortar joints should be correctly filled and should not exceed 6 mm in thickness.
(4) The amount of fall provided for the top surface of the sill should be 6 mm per 100 mm of sill surface.
(5) The projection of the sill beyond the brickwork face should be at least 50 mm.

Inspection of Sill

The sill should be inspected and examined to determine the cause of the deterioration.

Removal of Sill

Removing a badly defective sill should not cause any problems for the craftsman and it can be effected with the normal bricklayer's tools.

Cutting out should start at each end, taking out the first two bricks from each end; the third brick from the end on each side should be left to support the sill, and the bricks between should then be taken out. Temporary packings are then inserted, and the two supporting bricks are removed (figure 7.9).

After this operation has been completed the area of brickwork below the sill, termed the apron, should then be examined and, where repointing is necessary, raking out and cleaning should be carried out at this stage.

Reinstatement of Sill

The selected bricks should be placed on a flat surface and checked for alignment, with the correct joint allowance formed between the bricks; the length of brick can then be marked on the top surface, a gauge staff can be formed and the brick can then be cut to the required length (figure 7.10).

Having cut the bricks, dampen the bedding surface and where repointing is required dampen the area of the apron. Bed, joint and fix the first two bricks at each end, checking for fall, projection and gauge. A line can then be fixed to top or bottom arris, depending whether the sill is above or below eye level. The temporary packings should then be removed and the operation of bedding and fixing the brick on edge continued. This should be carried out by working from each end to the centre, checking for gauge as the work proceeds. The brick sill can be jointed after completion and the pointing of the apron also completed.

Where a wooden sub-sill is attached to the window the joint between the wooden sill and the brick sill should be pointed with a mastic compound.

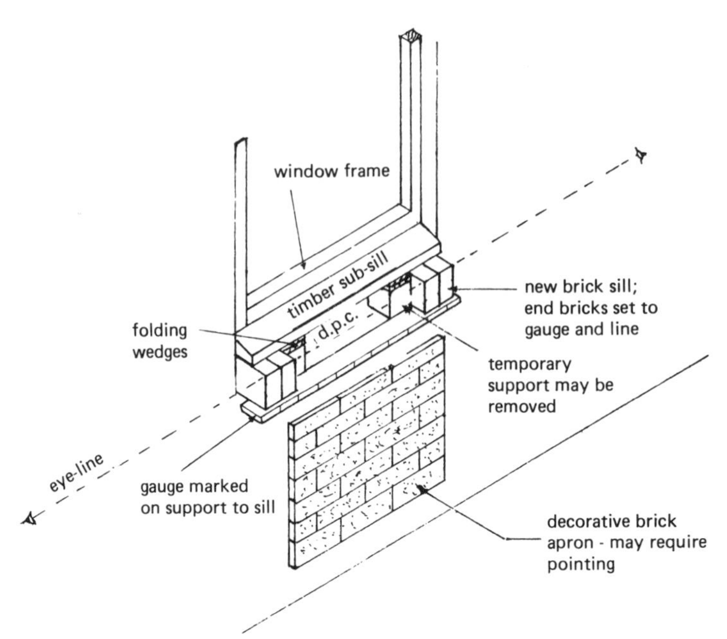

Figure 7.9 Method of renewing a defective brick-on-edge sill

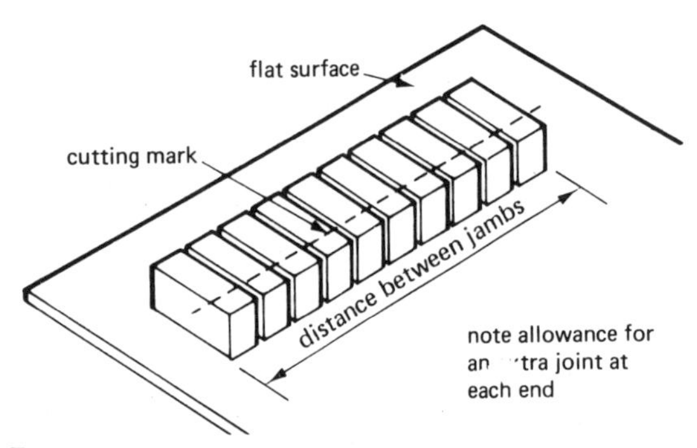

Figure 7.10 Setting out a brick sill ready for cutting

REPOINTING OLD BUILDINGS

Before any operations are carried out on the face of the building, it is important that a close inspection of the walling should be carried out.

Inspection of the Building Face

It is very important to determine the position of the walling and to recognise the amount of weathering or atmospheric pollution that the walling is required to withstand; the amount and type of deterioration of the wall surface, the number of laminated bricks and the condition of the mortar joints should all be carefully determined and assessed. Factors that should also be considered are as follows.

(1) Where staining of the brickwork face is evident, what is the type of staining, and how was the staining caused?
(2) Can the brickwork face be cleaned and what method of cleaning would be most effective?
(3) Is protection necessary, and how can it be ensured?
(4) Where bricks are defective, can they be removed and similar bricks obtained and then replaced?

It is very important that the above factors should be considered and a positive conclusion reached in each case before any form of work is started.

Scaffolding

Where scaffolding is required for defective brickwork to be repointed, cleaned and reinstated, careful consideration should be given to the type of scaffolding that will be most effective. It is important that scaffolding should be completely safe and should comply with the Construction Regulations. The main functions of scaffolding are as follows.

(1) Fatigue for operatives using the scaffolding should be reduced to a minimum.
(2) It should be possible for all work above ground level to be carried out in an economical manner.
(3) Cleaning operations should not be impeded and the scaffolding should provide protection for the work, and also for persons who are at ground level.

Protection

Protective measures should be carried out before any form of work starts on the building face. They may consist of masking out all existing mastic pointing, glass, paintwork, timber and decorative features. Masonry and ironwork should also be protected.

Cutting Out and Reinstating Defective Bricks

This operation should be begun at the top of the building and worked down to ground level. The bricks on each lift should be cut out and replaced before work is begun on the lift or platform below.

Repointing

Before this operation is begun it is often advisable, when large structures are involved, to point panels at the base of the walling. These are termed sample panels and they are used by the architect to determine the type of mortar that will be the most suitable for the building. Whenever possible sample panels should be at least 1.0 m^2 and labelled with

(1) the type of sand and cement used
(2) the cement–sand ratio
(3) the water content
(4) the type and amount of colouring used.

Panels should be pointed and viewed after a minimum period of 72 hours, otherwise a complete appreciation of the panel is not possible.

The type of pointing joint is also selected by the architect, and it is determined by

(1) the weather and degree of exposure
(2) the condition of the brickwork face
(3) the decorative requirements
(4) the need for economy in the pointing operations.

Obviously the sample panels are formed with the pointing joint selected, which then allows its qualities to be appreciated.

Cleaning Brickwork before Repointing

Brickwork staining is often caused by external sources although it can also be due to salts in the bricks or mortar. To remove stains on brickwork requires considerable knowledge, skill and care, otherwise the use of wrong techniques or materials may cause permanent damage to the entire face of the walling.

Removing Stains from Clay Brickwork

Oil Stains Sponge or poultice the area with white spirit, carbon tetrachloride or trichlorethylene. Where staining is severe, use several applications.

Efflorescence Allow weather to take its natural course, but brush off with a fibre brush when the efflorescence is at its maximum. After a reasonable period wash down each week for a period of one month and allow the walling to dry out before further wetting is contemplated.

Paint Stains Apply a patent paint remover as instructed, or use a solution of trisodium phosphate, 1 part to 5 parts of water (by weight). Allow the paint to soften and remove with a stiff fibre brush, washing down afterwards with soapy water.

Mortar Stains When possible use a softwood scraper and wash down with a diluted solution of hydrochloric acid, 1 part to 10 parts of water (by volume).

Lichens and Mosses First brush off with a stiff fibre brush, then use a patent moss killer as instructed, or a solution of zinc or magnesium silicofluoride, 1 part to 40 parts of water (by weight).

Rust or Iron Stains First wash down with a solution of oxalic acid, 1 part to 10 parts of water (by weight). If the brown staining does not respond it is probably a manganese stain.

Manganese Stains Brush down the staining with a solution of 1 part acetic acid, 1 part hydrogen peroxide and 6 parts water (by volume). Apply a second application only after a period of 3 days.

Lime Stains Treat as for mortar stains.

Smoke or Soot Stains Use a fibre brush and brush down gently. Apply a wash of household detergent, and where staining is heavy use trichlorethylene as a poultice.

Tar or Bitumen Stains Use a stiff fibre brush and scrub down with an emulsifying detergent. When the area is dry it may be necessary to apply a paraffin-soaked sponge.

All brickwork should be completely washed down whenever any forms of acid have been applied. This may cause efflorescence but it will only be shortlived.

Commercial Cleaning

This is only economical where large areas of walling are to be cleaned, otherwise it may be very expensive in labour and material costs. Methods used for stonework are steam, grit or sand-blasting, although the latter is only used as a last resort because of its effect on the surface of the walling material. Before steam, grit or sand-blasting is used, it is advisable to apply the methods on a sample panel before selection is made.

The above methods should only be considered for brickwork when it is judged impossible to obtain a clean surface with water and chemicals.

ALTERATION TO WALL LENGTHS AND THICKNESSES

Existing walls can be increased in length by any one of the following methods

(1) cutting out block indents
(2) cutting out toothings
(3) forming a slip or vertical joint.

Choice of Method

The method of extending the length of a wall is determined by the following factors

(1) the situation of the wall, that is, whether it is external or internal, and whether the walling is facework or commons
(2) the requirements of the designer
(3) the materials used for the existing wall
(4) whether differential movement is anticipated.

Methods

Block Indents

These are formed by cutting out indentations in the end of the existing wall. The indents should be at least 100 mm in depth and in blocks of odd numbers, that is, of three or five courses. The maximum depth of any block is five courses; with this method the indent accommodates an even number of tie bricks, two or four (figures 7.11–7.13)

When cutting out the indents, it is advisable to start at the top and work downwards, thereby preventing the tails of any bricks above from snapping off (figure 7.11).

Block indents are normally used to extend the length of internal brick walls where accurate and normal bonding arrangements are of secondary importance, but where adequate strength from bonding-in can be obtained.

Building-in Block Indents At the first course of the indent it is good practice to insert a wall tie or reinforcement (figure 7.13) to provide additional strength.

The indent is then built up with considerable care being taken to caulk up the top bed joint in the indent with semi-stiff mortar. Obviously, if the existing walling is dry or dusty it should be brushed down and damped before joining up takes place.

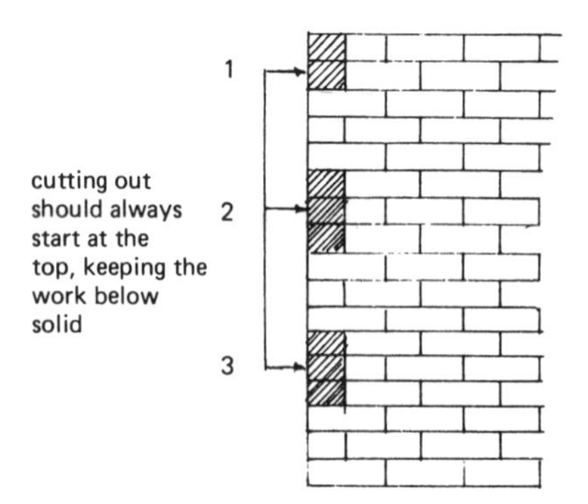

Figure 7.11 Preparing for block indents

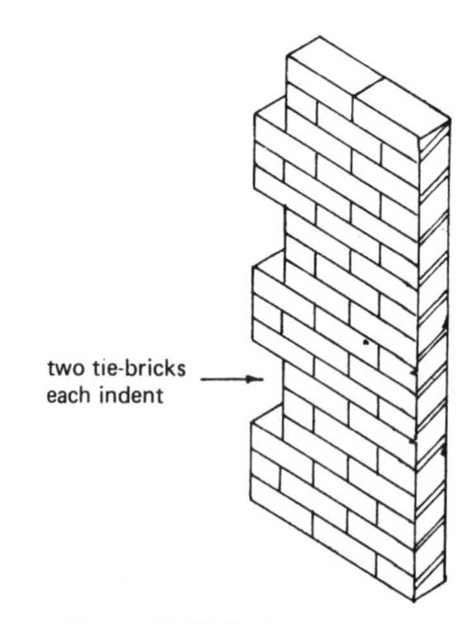

Figure 7.12 Indents cut out

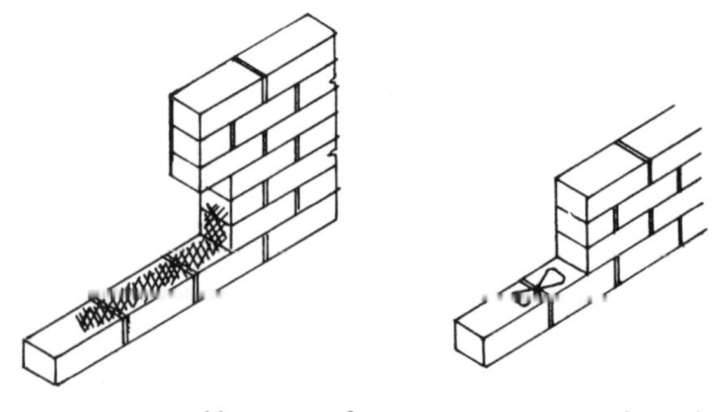

Figure 7.13 Using reinforcement to strengthen the tie when increasing wall lengths with block indents

Toothings

This method consists of cutting out every alternate brick at the end of the existing wall. The depth of the indent or toothing should be 56 or 110 mm, depending on the bonding arrangement of the existing wall and the new walling to be attached. Again it is essential to start cutting out the indents at the top and to work downwards, to prevent breaking off the projecting tails of the existing bricks (figure 7.14). When

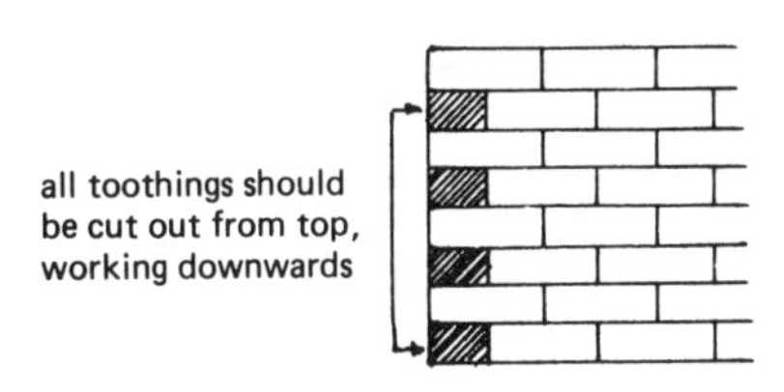

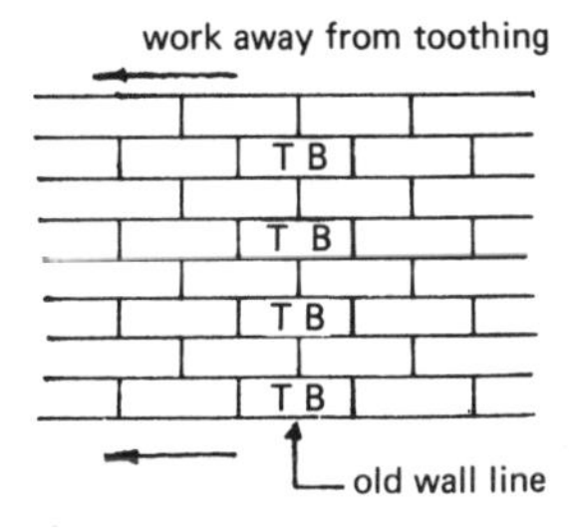

Figure 7.14 Increasing wall lengths by the toothing method

cutting out is completed, the toothings should be brushed out and if they are dry, damping will be required to ensure good adhesion between mortar and bricks.

The toothing method is extensively used on all external facework or wherever the bonding arrangement must be seen to be continuous throughout the length of walling (figure 7.14).

Building-in Toothings Before any building work is started it is essential to ensure that the following recommendations are carried out.

(1) The dimensions of the new bricks should be the same as those used in the existing wall.
(2) The mortar should be the same density, texture and colour as the mortar in the existing wall.
(3) The toothed ends on the existing walls should be checked for vertical alignment, thus avoiding cutting to provide the correct bonding arrangement.

When building-in to toothings, line and pins should always be used to ensure horizontal alignment. Each tie brick must always be inserted first and the course of bricks continued from this. The tie brick should never be inserted last, otherwise caulking and pinning up cannot be carried out satisfactorily. The caulking up should be completely solid and carried out with a semi-stiff mortar.

Slip Joint

This is sometimes referred to as a butt joint. The method consists of simply forming a vertical mortar joint, minimum 12 mm, at the end of the existing wall and starting the new work from this position. Often an open joint can be formed or butyl rubber or polysulphide compounds can be inserted. Both can have a mortar joint applied on the face later but the former method of leaving an open joint requires a movement joint to be gunned-in when the walls are completed (figures 7.15 and 7.16).

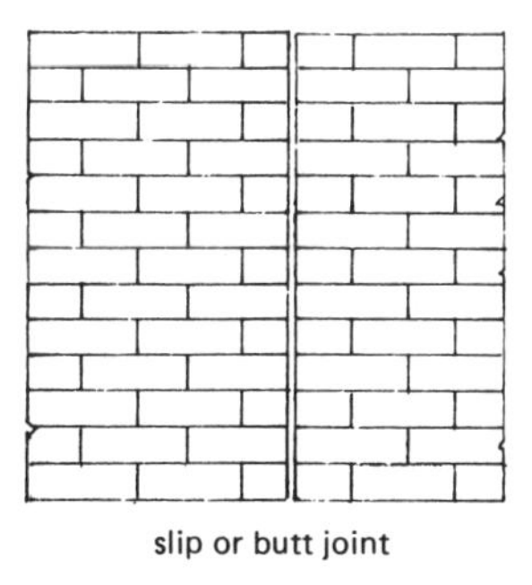

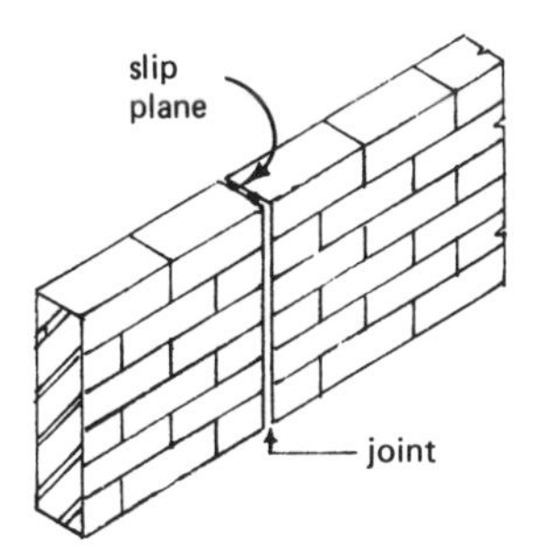

Figure 7.15 The slip or butt joint used in brickwork

Slip joints are used where there is little knowledge of the existing foundation structure, or where unequal settlement is a possibility and differential movement may be expected.

It is important that coursing through from the existing brickwork is carried out. All the new work should again start from the vertical joint position to the opposite end of the new wall.

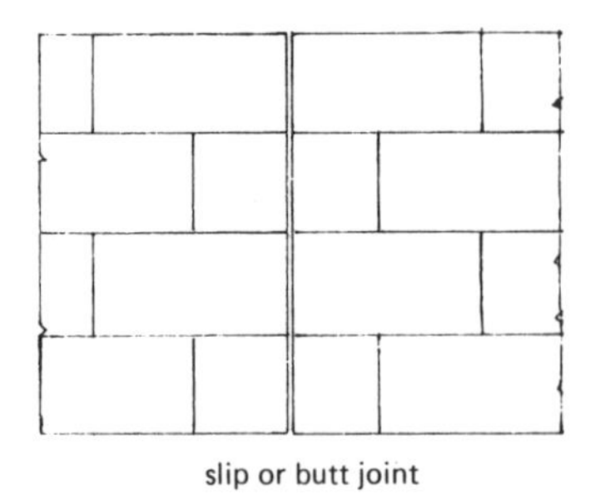

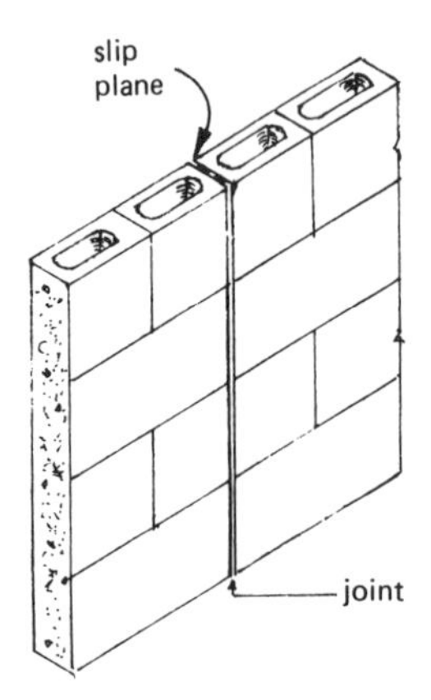

Figure 7.16 The slip or butt joint used in blockwork

Treatment for Extending Block Wall in Length

Block walls can be increased in length by any of the three methods previously mentioned, but where indents are to be cut into the end of the existing block wall, it may be easier, more accurate and economical to use a portable electric saw with a masonry blade or disc. This method does not affect the stability of the existing wall as often happens when heavy percussion tools are used. The procedure for building-in the blockwork is the same as for brickwork; the mortar should only be as dense as the walling blocks being used. The blocks used to form the new walling should be the same as those used for the existing wall; where this is not possible, joining up should be carried out with the slip joint.

INCREASING THE THICKNESS OF EXISTING WALLS

Existing walls can be increased in thickness by block bonding. This is considered the most practical method of tying two walls together to increase the thickness and also obtain the maximum amount of stability.

The method consists of cutting indents or recesses into the face of the existing wall; the indents should not exceed 100 mm in depth, while the length should be 225 mm and the height of the recess should be three courses (figures 7.17). The indents should be placed diagonally at 45° to the horizontal and cover the entire face of the wall (figure 7.18). Cutting out is performed with the normal bricklaying tools and also the portable electric saw with a masonry disc. Before building operations are begun, all indents should be brushed completely free of dust and damping down should be carried out when necessary.

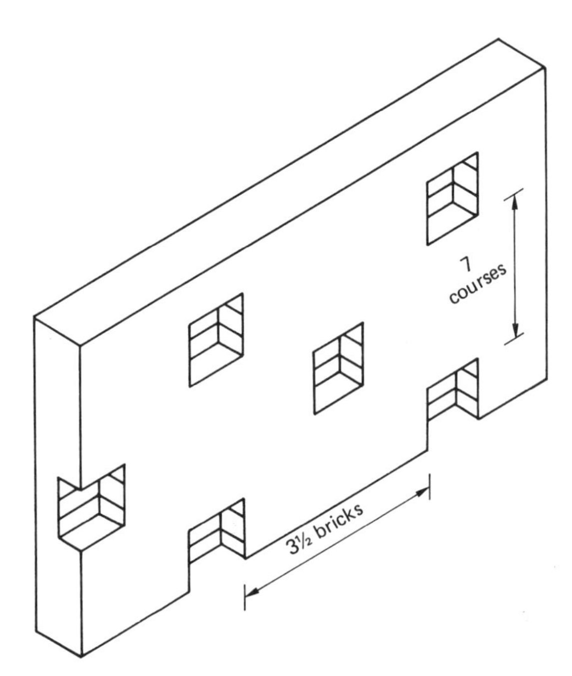

Figure 7.17 Forming block indents for increasing the thickness of a wall in Flemish bond

Building Up the New Wall

Considerable care should be taken in the setting-out operations. The bond for the new facing wall should be arranged to ensure that header-tie bricks occur in each block indent. Whenever English bond is used for the facing wall there should be four tie headers per indent (figures 7.18 and 7.19). Flemish bond permits only two tie headers per indent (figure 7.20).

Caulking up at the top of each indent should take place to ensure the completion of the tie. It is good practice to use horizontal reinforcement in the new facing wall; wall ties should be used in all block indents.

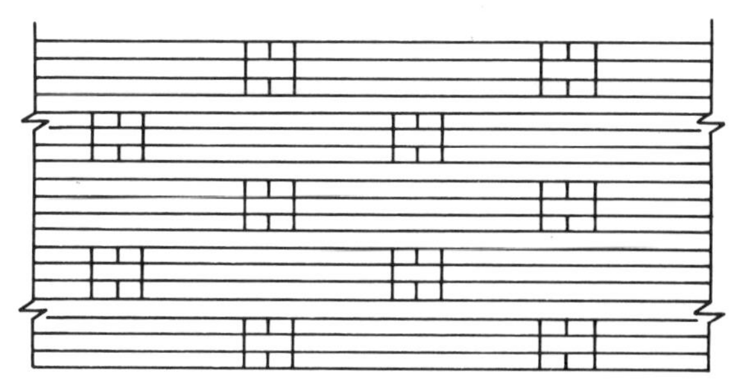

Figure 7.18

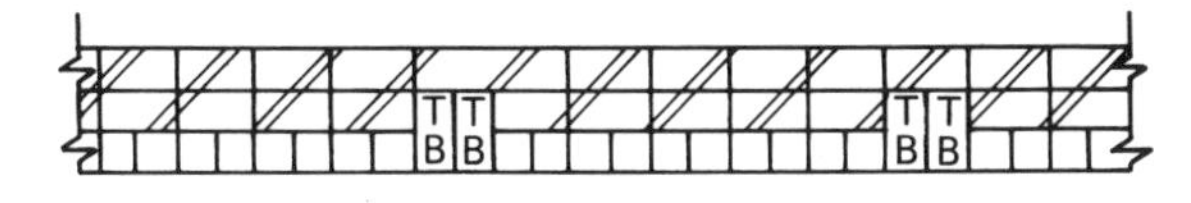

Figure 7.19

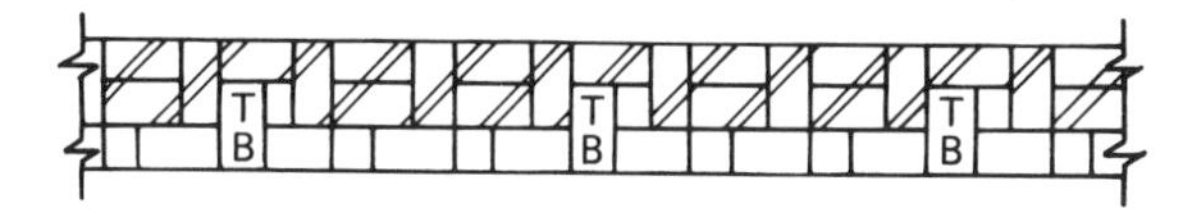

Figure 7.20

FIXING CANTILEVER BRACKETS

Cantilever brackets can be fixed to solid or cavity walls of bricks or blocks. Before any operations are started it is necessary to inspect the walling because this often determines the types of tool to be used, the materials required and the type of temporary fixing equipment necessary.

It is important to recognise the condition of the walling and to determine whether it is of solid or cavity construction.

Procedure for Fixing a Line of Four Cantilever Brackets

First check whether the floor is level because this may alter the height of the brackets, depending on the designer's fixing requirements. A line should then be formed on the face of the wall at the required height. This is done with a chalk line, level and possibly a straightedge. The position of each bracket is then marked at centres on the chalk line.

Cutting-out operations can now be started. The cantilever brackets must be inserted into the wall to a depth of at least 100 mm to become effective, therefore it is essential to obtain a clean cut when cutting to form the soffit at the top of the bracket because upward movement cannot be allowed.

The dimensions of the holes should be not more than 28 mm on either side of the bracket. Whenever possible the bottom of the bracket should rest on the brickwork, which will ensure improved stability for the bracket and make fixing easier.

An assembled timber jig should now be set up, either by the carpenter and joiner, or the bricklayer craftsman himself. The jig should be adjustable to ensure that it is correct and level; the top rail should also be adjustable to ensure a level line for the brackets, therefore it is necessary to use folding wedges in both positions, as in figure 7.21. The jig should finally be positioned for alignment of the end of the brackets. After the timber jig has been set up, the brackets should be temporarily inserted and checked for height, level and alignment. When this operation has been completed the brackets should be removed and the holes for the brackets then damped only sufficiently to ensure adhesion.

Mortar for making good around the end of the brackets should be cement and sand, reasonably dense, with a ratio of 1:3 or 1:4, although it should not exceed the density of the walling material. The first bracket is then fixed into position, with the square shank end inserted into the hole formed in the wall face, and around the shank the semi-stiff cement mortar can now be compacted. If the hole is overlarge, small particles of bricks can be inserted. Where the walling material is of hollow blocks, then before fixing is carried out, the hollow portions of the blocks should be made solid; this operation should be performed at least 24 hours before the actual fixing of the brackets.

After all the brackets have been fixed, a further check should be carried out for level and alignment. The fixing equipment should not be removed until at least 48 hours after fixing. Care should be taken not to disturb the brackets, therefore the folding wedges will facilitate the easing operation.

The same procedure can be used for fixing a single bracket, but only a single prop is required to provide support (figure 7.22).

FIXING RAG BOLTS

Rag bolts, often termed holding-down bolts, are normally hand made from wrought iron. They are formed with a ragged base and sides to prevent displacement and ensure complete security in the flooring material. The tops of the bolts are circular, with turned threads to receive washers and a threaded nut (figures 7.23 and 7.24).

Rag bolts are used extensively when heavy, moving machinery is to be anchored to concrete floors. The

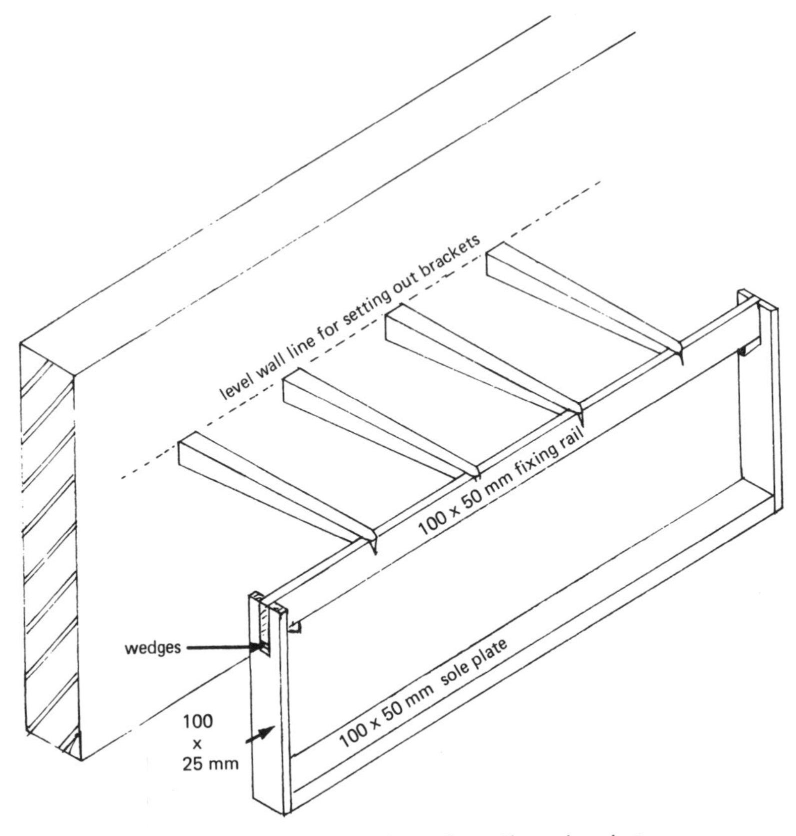

Figure 7.21 Method of fixing a line of cantilever brackets

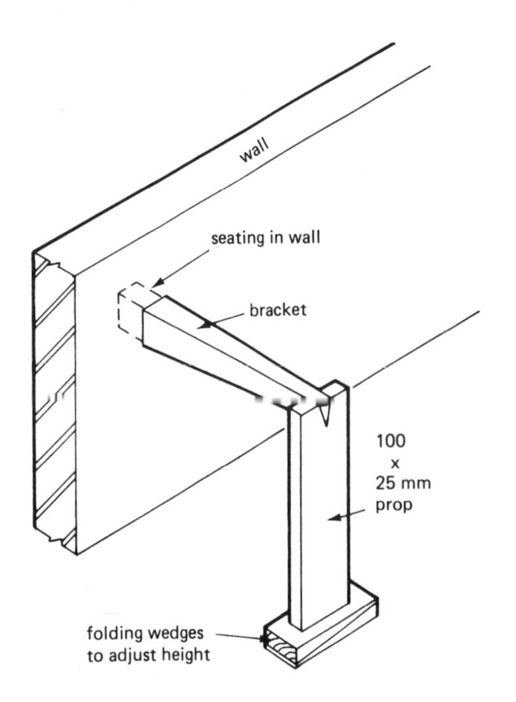

Figure 7.22 Fixing a single cantilever bracket

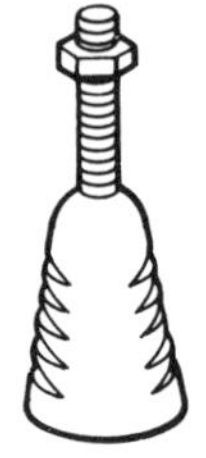

Figure 7.23 Rag bolt with nut and straight base

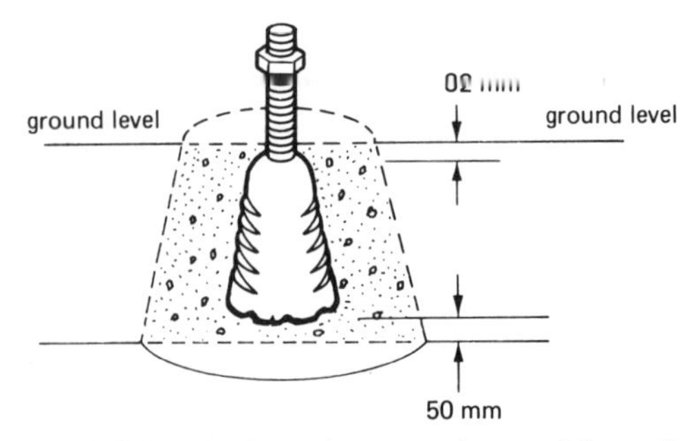

Figure 7.24 Rag bolt with nut and ragged base fixed in hole and surrounded by concrete

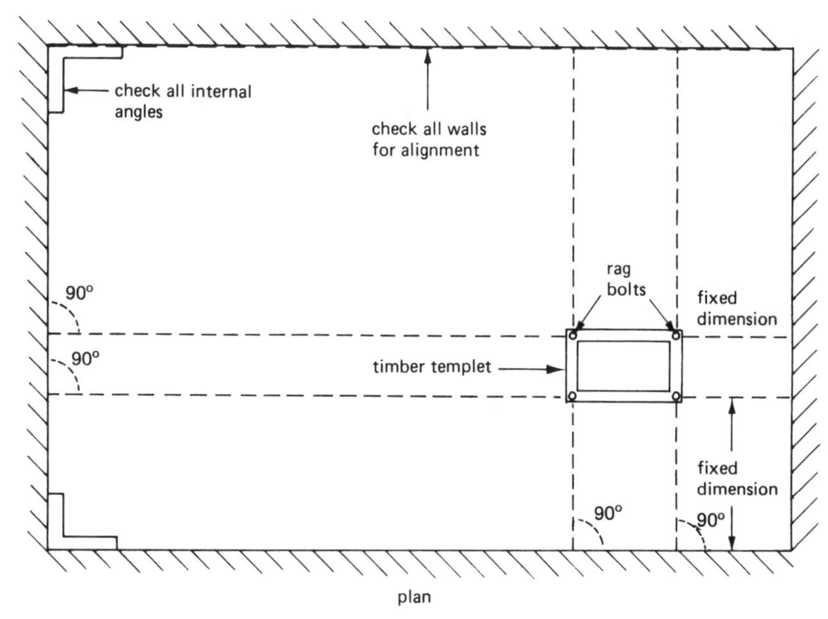

Figure 7.25 Methods of setting out for rag bolts

diameter of the top of the bolt is formed to fit into the hole or slot at the base of the machine.

Setting Out

It is important that considerable care is exercised at the setting-out stage, before the rag bolts are fixed. Measurements should be checked, levels taken, squareness ascertained and, where there is any doubt, checks should be made again. It cannot be overstressed that accuracy is of the highest importance since any error could prevent the efficiency and functioning of the machinery. It is obviously very necessary to construct a timber templet, which can be used to aid the setting out and to assist with fixing operations (figure 7.25).

Hole drilling should be carried out with a lightweight Kango, but where the concrete is excessively thick and very dense, it is advisable to use a heavy Kango. The use of the lump hammer and cold-steel chisel, although useful for drilling and cutting a single hole, would be uneconomical where several holes are required.

After the holes are formed, cleaning out is then required (figure 7.26) and the sides of the concrete should be adequately damped. The timber templet, which positions and supports the bolts, is then placed according to the engineer's drawings (figure 7.27).

The bolts are suspended from the templet in the holes below and a concrete mix of density equal to that of the existing concrete floor is then made. Where the diameter of the holes is less than 100 mm the aggregate size for the concrete should not exceed 12 mm but 18 mm aggregate can be used for larger holes. The water content of the concrete should be reduced to produce a minimum of laitance. The concrete is then placed around the bolt and compacted with a 19 mm-diameter rod and the surface is finished

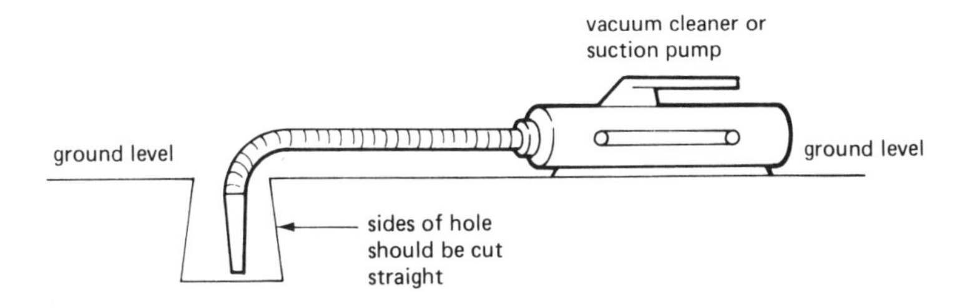

Figure 7.26 Method of cleaning hole after cutting

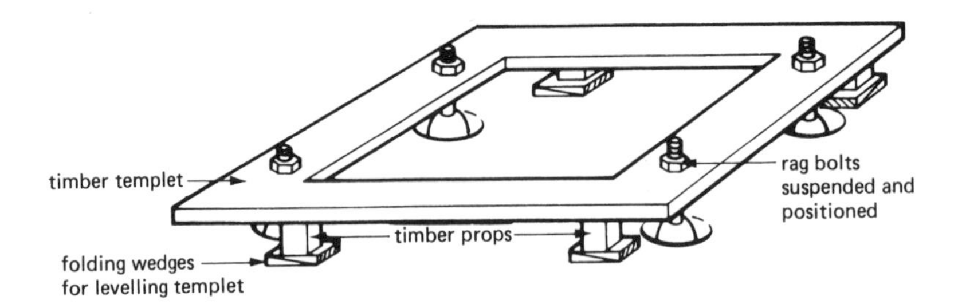

Figure 7.27 Adjusting and positioning rag bolts

off with the steel float. It is important to cure the concrete around the bolts and a minimum period of 72 hours should be allowed.

The removal of the timber templet should be undertaken with considerable care. The nuts are removed and the templet gently lifted from the bolts. Threads should be checked for cleanliness and where necessary covered with a suitable grease; before leaving, it is again advisable to carry out another check on all the bolts for level, position and dimensions.

8
PAVING

Europeans are extremely fortunate in being able to enjoy and appreciate areas of paving, which have been their heritage for many centuries.

Paved surfaces allow the pedestrian to move with the minimum of fatigue, they complement and enhance buildings and decorative features, and have stood the test of time when used for roads. It is an accepted fact that, in comparison with other forms of construction, the merits of the paved surface are the least recognised. The skills of the designer and craftsman do not receive the appreciation that they deserve. This is possibly because the pedestrian accepts and is familiar with the paved surface. It is to be hoped that this trend is now being reversed and that people are becoming more aware of the aesthetic qualities and importance of the paved surface.

Before considering the type of paved surface that is to be constructed, it is necessary to obtain the following information

(1) the total area of paving required
(2) whether the situation is external or internal
(3) the type of traffic expected
(4) the decorative requirements
(5) the amount of abrasive resistance required for the surface
(6) the requirements for removal of surface water.

When the requirements have been determined, the types of material can then be selected. Paved areas are normally formed with the following materials: concrete slabs, stone slabs and paving bricks.

TYPES OF PAVING SLAB

Concrete Paving Slabs

These are obtainable in two forms, the pressed, vibrated and reinforced slab and the pressed, unreinforced slab. Concrete slabs can have smooth or non-slip surfaces. They are produced in sizes varying from 600 x 600 x 50 mm to 900 x 600 x 75 mm, although it is possible to obtain smaller slabs of 300 x 300 x38 mm.

The usual method of cutting unreinforced concrete slabs is with the hammer and chisel or portable electric saw, but the need to cut *reinforced* slabs should always be avoided because failure is almost certain owing to the reinforcement.

Stone Slabs

These are usually made from sandstone or Yorkstone and they are produced with sawn surfaces. The sizes of stone slabs vary, but for normal highway construction 900 x 600 x 75 mm slabs are used. Stone paving slabs are now considerably more expensive than concrete slabs but they are often more resistant to abrasion and possess more decorative qualities. A disadvantage with stone slabs is that they easily become stained, and constantly increasing costs tend to preclude their use, especially for public footpaths; as renewal becomes necessary the concrete slab is usually used as replacement.

Stone slabs are usually cut with the hammer and pitching chisels. Before laying and fixing any type of paving it is necessary to determine the amount of fall or slope required to remove surface water. This is normally between 1 in 40 to 1 in 60, but the amount of fall provided should not increase the physical effort of walking. The direction of traffic is required to determine the direction of the joints in the paving (figure 8.1).

Laying Concrete Slabs

These are usually bedded on a sub-base of sieved clinker ash or sand and placed on mortar dabs under the corners and centre of the slab. The thickness of the mortar dabs forming the bed should not exceed 32 mm and joints between slabs should not exceed 6 mm. The mortar used for bedding is 1:6 cement–sand or 1:2:6 cement–lime–sand (figures 8.1 and 8.2). The same bedding treatment and joints are used for stone slabs, and the mortar mix is also the same.

Concrete and stone slabs should be jointed before the end of the day's work and with the same ratio of mix as for bedding.

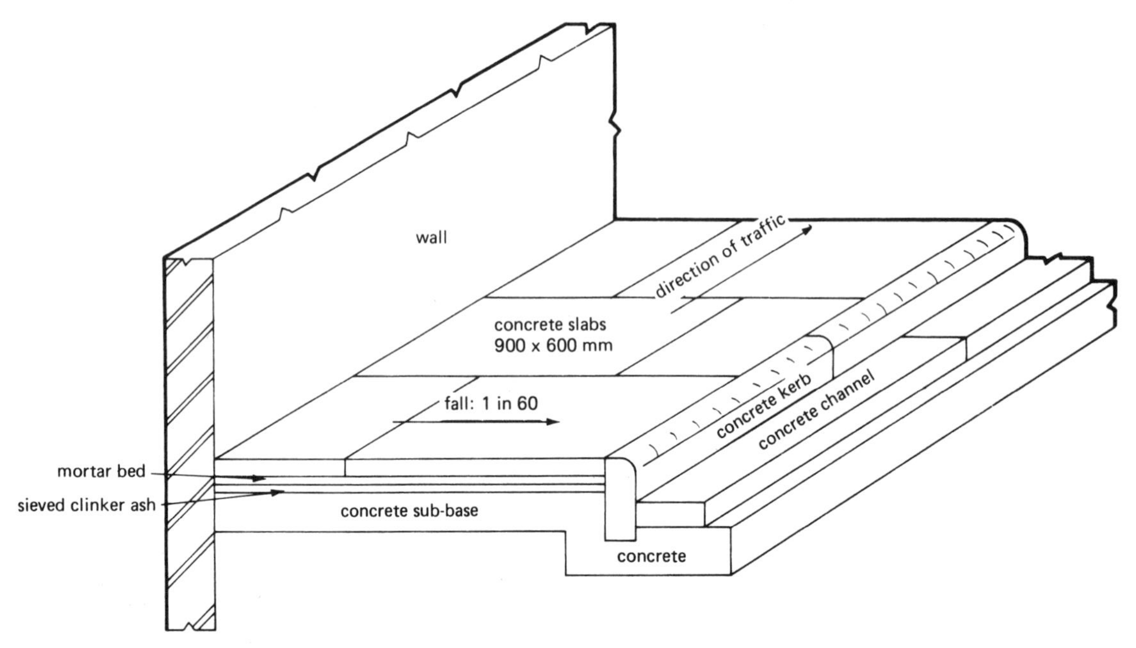

Figure 8.1 Paving slabs on mortar, sieved clinker ash and concrete sub-base

Treatment of Concrete and Stone Slabs

When stacking and storing concrete or stone slabs it is necessary to stack them on edge to ensure complete dryness of the slabs, therefore, whenever possible, these paving materials should be covered with lightweight sheeting.

Tools and Equipment for Laying Paving Slabs

The hardcore base is usually positioned with normal excavating tools – shovels, picks and hammer – but for consolidation a rammer or punner is necessary. These are hand or mechanically operated.

Slabs are usually laid with the beedle or mawl,

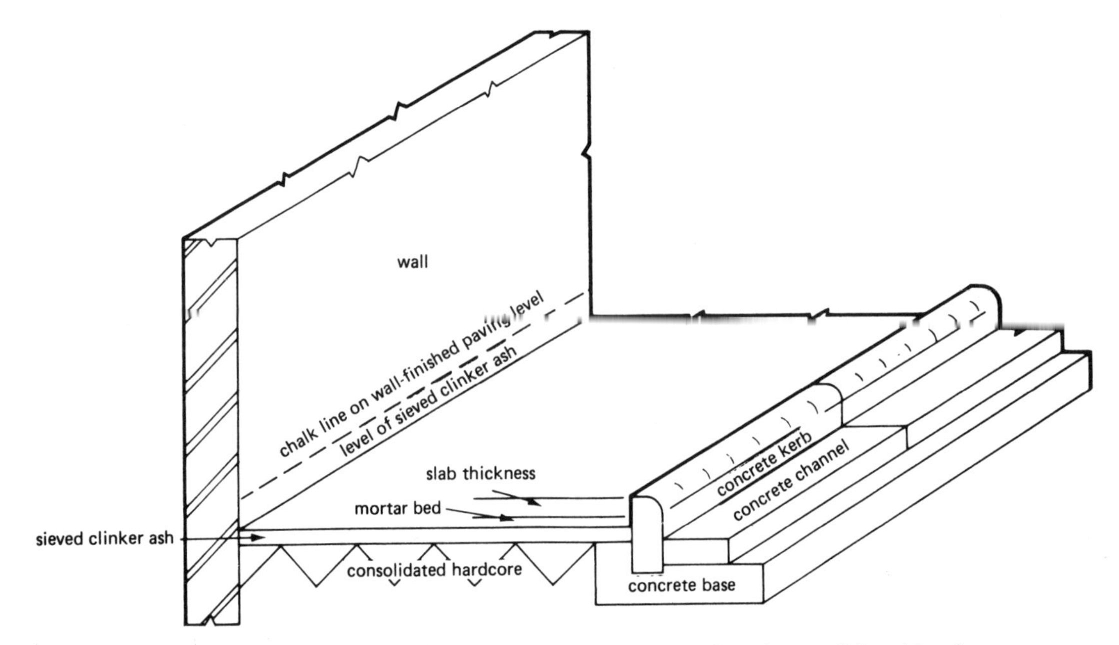

Figure 8.2 Method of laying slabs on sieved clinker ash and consolidated hardcore

which is a large rubber-headed hammer. Paved areas are normally levelled with the aid of the Cowley level and wooden pegs, although boning rods are often used. Straightedges used for checking the slabs are often tapered to the amount of fall required (figure 8.3).

BRICK PAVING

This is the most attractive and decorative form of paving. Because of the great variety of clays found in European countries, a considerable variation in colour and texture can be obtained. The flexibility of bonding arrangements allows the geometrical patterns to enhance the colours of the paving. Brick paving can be used for industrial flooring, where the qualities of the paviors are used to combat abrasive wear and tear, or for domestic use, either external or internal, and also where a decorative appearance is necessary to provide the aesthetic qualities required by the architect.

Brick paving is formed with bricks laid on edge or flat; although the pressed brick is obviously the better type of pavior, wire-cut bricks are often used laid on edge or flat.

For industrial use engineering bricks or paviors are necessary. The surface of special paviors, produced to withstand very abrasive wear and acids, and also to provide a non-slip surface, is often chequered or impregnated to form a dimpled pattern. This type of brick should be produced to meet the requirements of BS 3679. When cutting is required an abrasive wheel, that is, a brick saw, should be used.

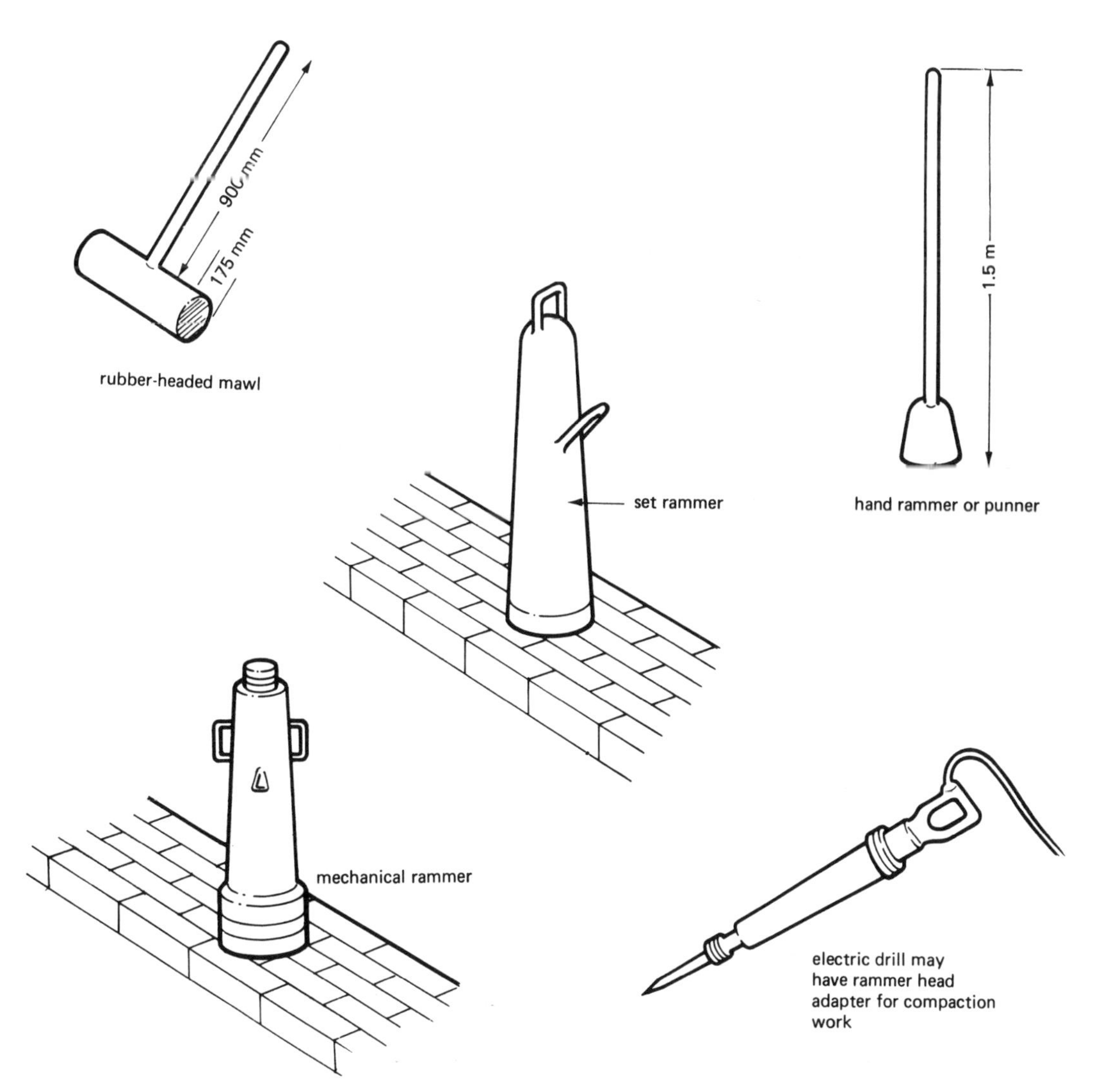

Figure 8.3 Compaction and consolidation equipment used in paving work

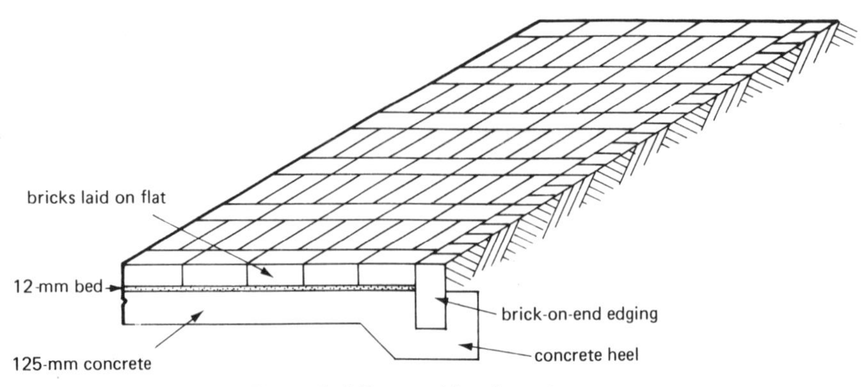

Figure 8.4 External brick paving

Brick Slips

These are often used to provide paved surfaces. The size of slip varies from 215 x 65 mm, with thicknesses of 25, 33, 40 or 50 mm. Slips can be used to form industrial or domestic surfaces, but, whatever the purpose, they should be treated with a different laying and bedding technique from that used for normal brick paving.

Laying and Fixing Brick Paving

Before starting to lay brick paving, it is necessary to consider the following factors

(1) the paving material
(2) the type of base
(3) the form of bedding
(4) the mortar joints

Paving bricks can be laid on a mortar bed, on compacted damp sand or in bituminous compounds. The types of brick, the situation and the use of the surface obviously influence the type of base required. For industrial internal paving, the base should be of concrete, with a thickness of between 100 and 150 mm, with a ratio of 1:2:4.

External footpaths require a 100 mm thickness of concrete, while garden paths may have sufficient strength with a concrete base of 75 mm, provided that a consolidated layer of hardcore is formed below the concrete (figure 8.4).

Internal paving formed on a concrete base requires protection from rising damp, therefore it is essential to provide a damp-proof membrane on the surface of the concrete base. This separating layer can consist of polythene sheeting, building paper or bituminous felt. These should all be provided with lapped joints where necessary, and be laid on completely flat surfaces, but should not be provided with any form of adhesive bonding with the concrete (figure 8.5).

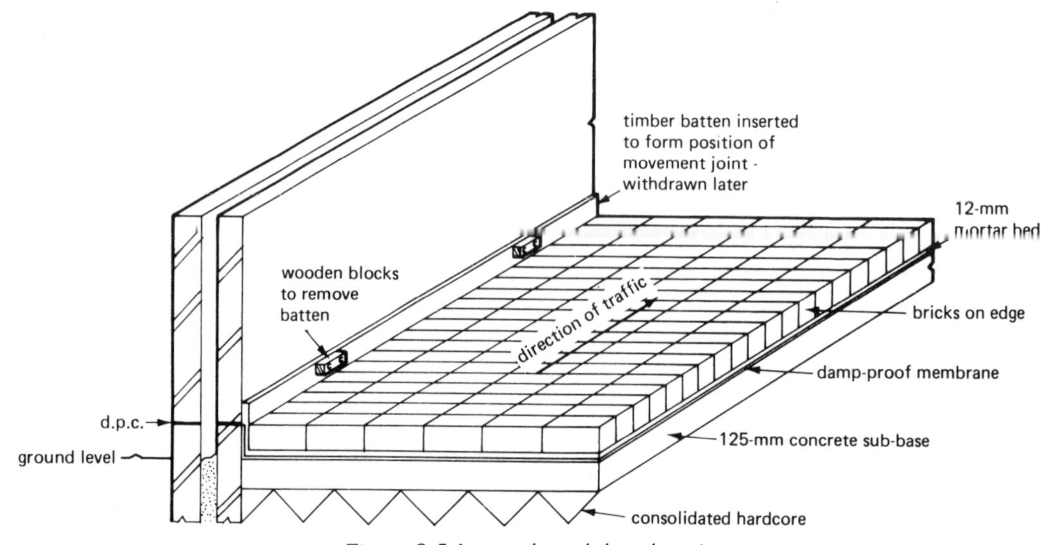

Figure 8.5 Internal stack bond paving

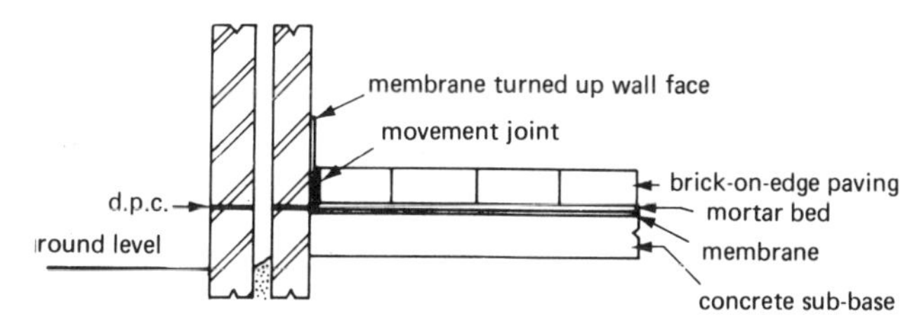

Figure 8.6 Membrane formed behind movement joint

Wherever the concrete base forms contact with external walling, the damp-proof course in the wall and the damp-proof membrane should be adequately lapped to prevent moisture penetration (figures 8.5 and 8.6).

When laying the concrete base it is advisable to provide the required amount of fall within the thickness of the concrete, which, as previously mentioned, may be 1:40, 1:50 or 1:60, depending on the amount of water expected on the surface of the paving.

The mortar bed for the paving should always contain a minimum amount of water. The mortar mix varies according to the types of brick or pavior used, and may be 1 part cement to 4–6 parts clean building sand, with ¼–½ part lime where required. Plasticiser is not advisable for mortar bedding paviors.

The bed should be at least 10 mm and not more than 20 mm thick. For all types of brick paving the beds should be constructed with a straightedge by forming mortar screeds (figure 8.7). The bricks should not be placed on the bed until all laitance has been removed from the mortar surface.

Laying and Forming Joints

Whenever possible, it is good practice to use working lines when laying brick paving. Where the bonding arrangements provide courses, the working lines should be used in the normal way (figure 8.8). Where sectional bonding is formed it may be possible to use the lines at the positions of each section of unit. Most bricks have a mortar joint appended before laying, and this should be kept at least 12 mm from the top surface to allow jointing to take place later.

Mortar joints can vary from 6 to 10 mm in thickness. After the bricks are laid, they should be lightly tamped with a wooden beating block, then checked for alignment with working lines and straightedge before jointing takes place. To provide the required amount of adhesion, all clay bricks should be lightly damped before using, but engineering paviors and brick slips should be perfectly dry for laying.

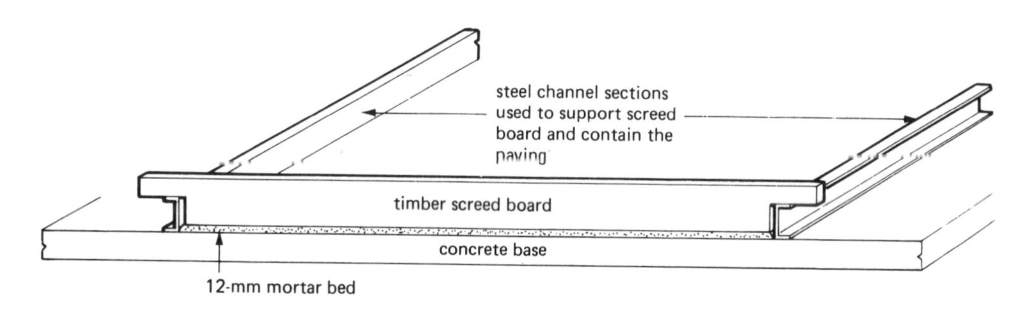

Figure 8.7 Method of forming mortar bed for paving

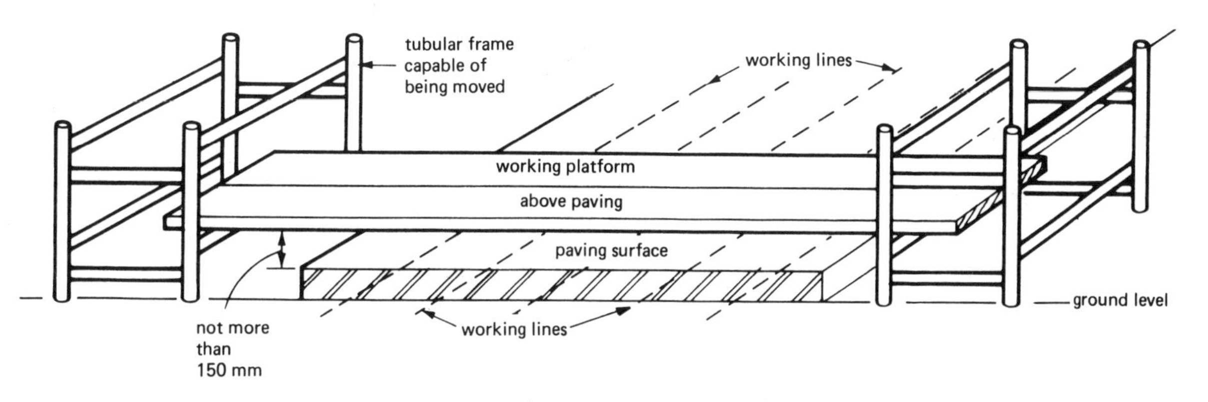

Figure 8.8 Method of working for laying paving

Jointing

This should be carried out after the required area has been paved, but during the same day. The mix ratio for the jointing mortar should be the same as for the bedding mortar, but with a minimum water content, and the mortar should be semi-stiff.

Another method of jointing used for paved surfaces is to use a semi-dry mix, and gently brush over the surface of the paving with a fibre brush. The disadvantage of this method is that compaction of the joints may not be complete, also staining of the surface can occur.

Direction of Joints

When designing the paving the direction of the joints should always be considered. When the paving is required for industrial use and to withstand abrasion the paving units should always be bonded and longitudinal joints should be eliminated. The transverse joints should always be at right-angles to the direction of the traffic (figure 8.5).

Movement Joints

These should always be inserted where the sub-base and the paved surface abut any walling, piers, columns or machinery, or when the paved area exceeds 6.0 m in any direction. Obviously, the size of the bay will determine the amount of movement, which will also be influenced by thermal activity and the movement of the paving materials themselves.

To accommodate movement, a joint should be formed around the perimeter between the paving and walling by the insertion of timber battens. These are withdrawn before the curing stage and a movement joint is inserted. The materials used to accommodate movement in paving are butyl rubber, polyurethane and silicone rubbers. Where necessary the surface of the joint can be protected by forming a sealing joint of polysulphide compound (figure 8.9).

Laying Techniques

Paving work should only be performed when other craftsmen have completed their operations. It should be programmed as a finishing operation and only the actual decorative finishing should be performed after the paving work. During the entire operation of paving it is essential for the craftsman to work above the level of the paving, that is, while he is laying the bed, placing and fixing the bricks, and also tamping and jointing. With this method the craftsman is never in a position where he may disturb the bedding or cause any misalignment of the paving (figure 8.8).

Curing Paved Surfaces

After the jointing operation has been completed, the finished work should be allowed to mature. This is assisted by curing. Where it is deemed necessary the floor surface can be protected with light timber battens and polythene sheeting. After a period of 24 hours, the surface can be damped by applying a fine spray of water. This treatment should be continued for a further 72 hours and, for complete maturity, the surface should be closed to all traffic for another 48 hours.

Brick Paving Patterns

Patterns for paved areas are usually determined by the following

(1) the area involved
(2) the surface resistance required
(3) the decorative requirements.

While the same bonding arrangements are obtained with both bricks on edge and bricks laid flat, the decorative appearance in each case may be quite different. The brick on edge contains more joints and yet possibly provides a stronger surface area. The patterns should always be arranged to enhance the area involved. A Flemish or stretcher bond arrange-

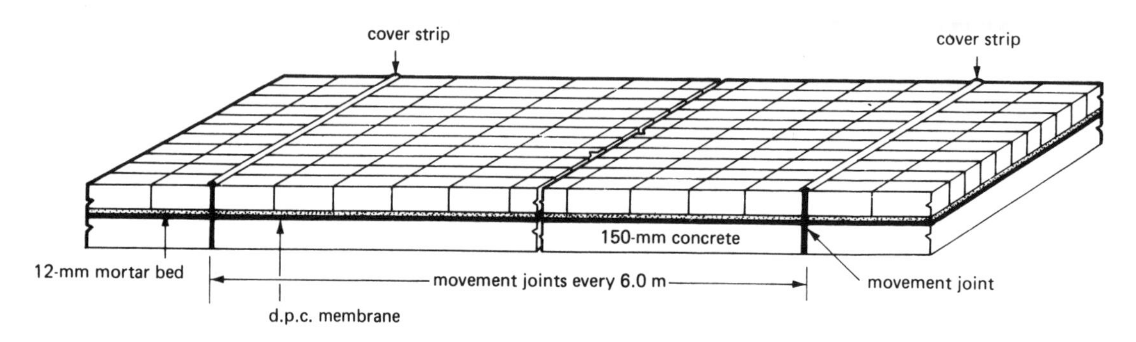

Figure 8.9 Position of movement joints

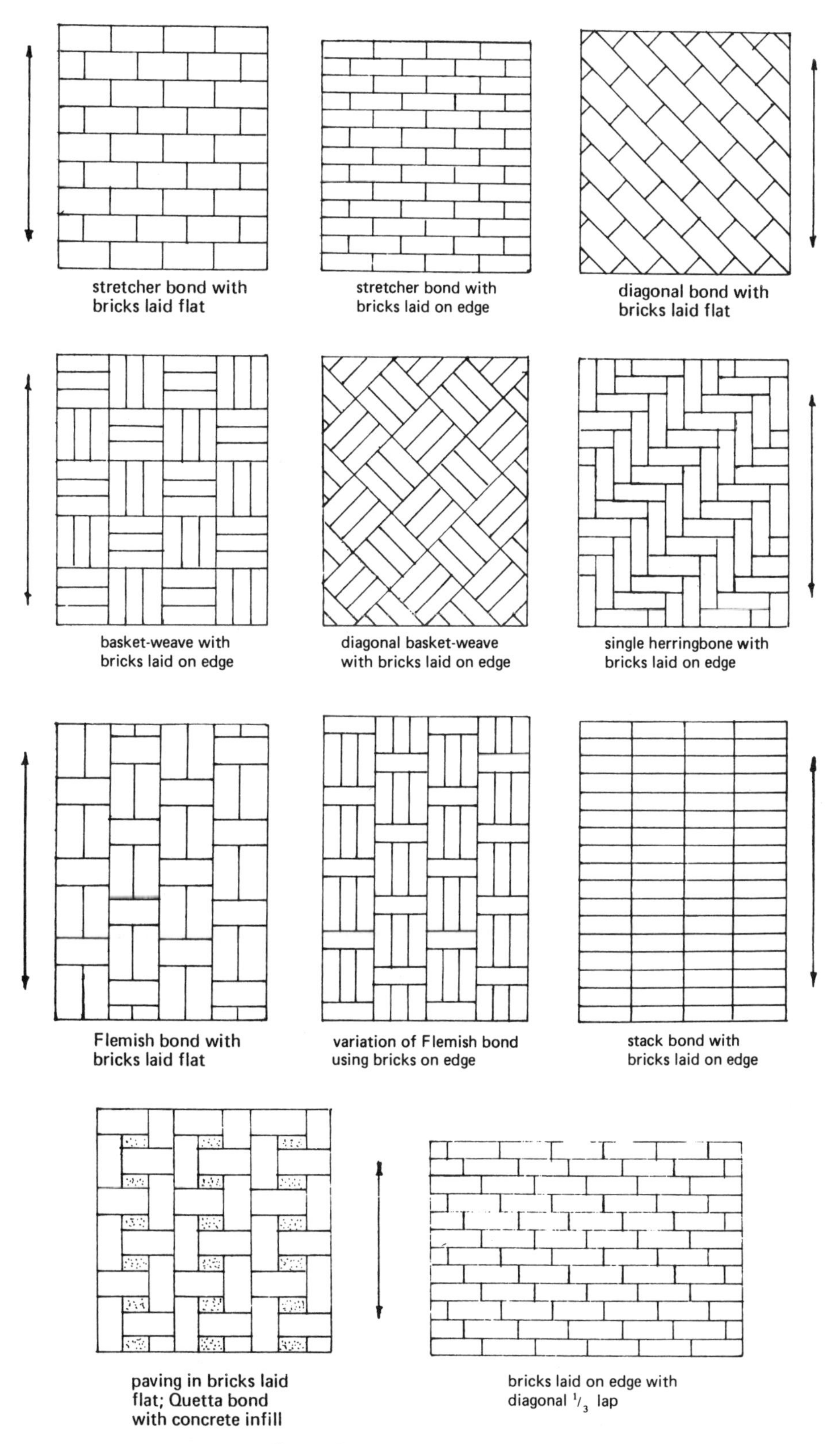

Figure 8.10 Brick paving patterns

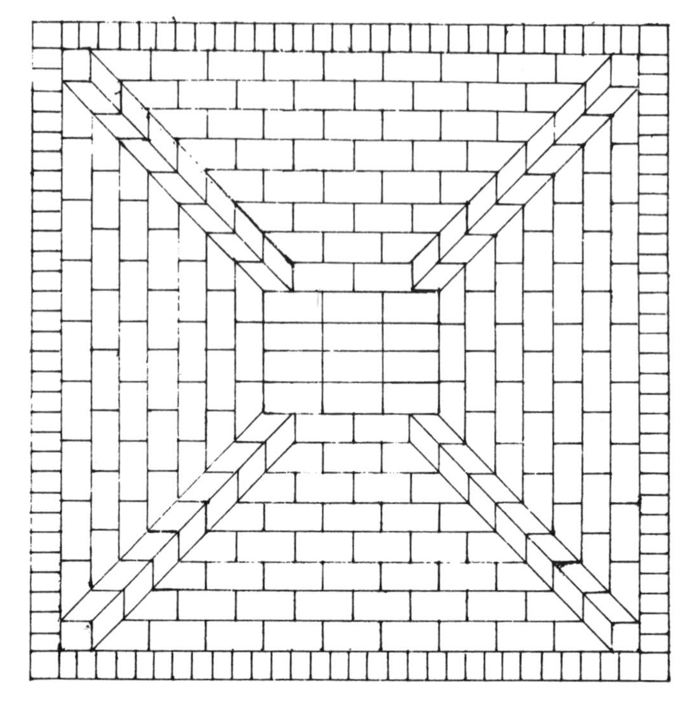

Figure 8.11 Decorative brick paving with bricks laid flat with brick on end to form edging

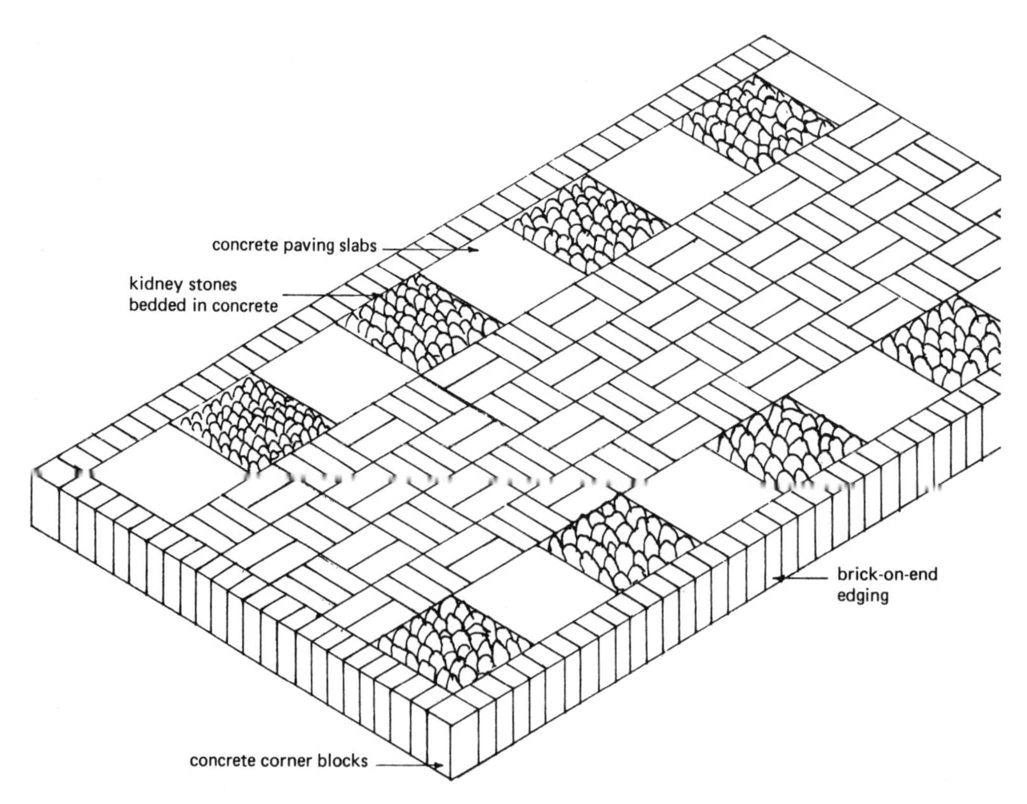

Figure 8.12 Compound paving used externally

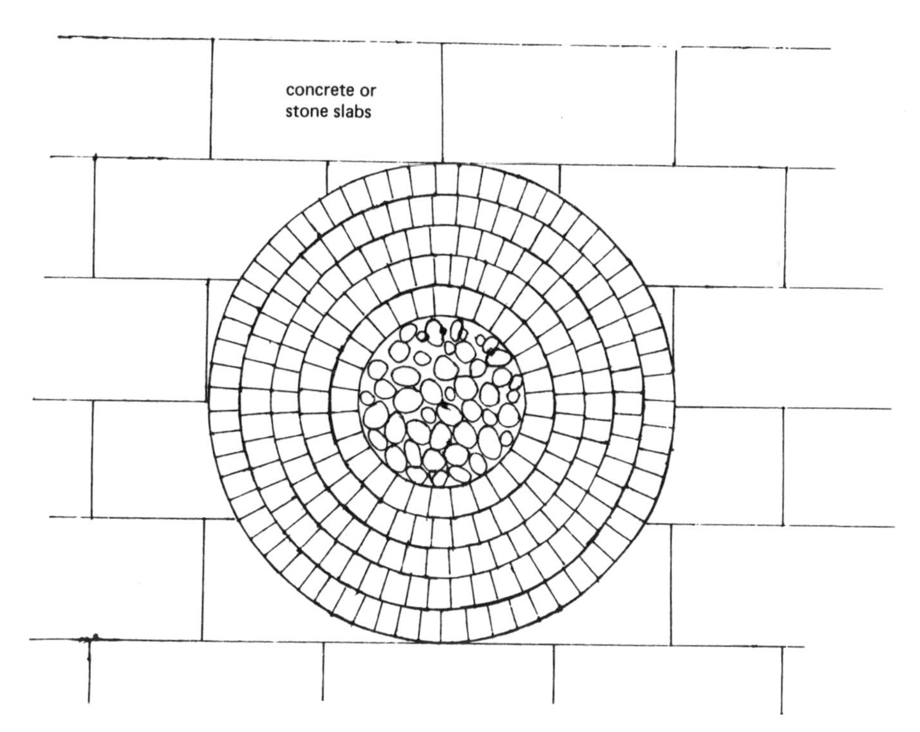

Figure 8.13 Circular brick paving used as a decorative feature. The bricks are laid on end with kidney stones or exposed aggregate as infill

ment is often suitable for large areas of paving, while narrow areas are enhanced by a diagonal or herringbone arrangement. When setting out the bonding pattern it is essential to eliminate the amount of brick cutting, therefore rectangular areas should normally be set out from a base line, and square areas from the centre. Both methods will normally ensure that cut bricks, where required, will be the same size and occur around the perimeter of the area. It may also be advisable to adjust the thickness of joints because on large areas this can eliminate a considerable amount of cutting (figures 8.10 and 8.11).

COMPOUND PAVING

Paved areas formed with more than one material are termed compound paving. Designers are now using combinations of bricks, concrete slabs and often flints and kidney stones (figures 8.12 and 8.13). When the craftsman is required to lay areas of compound paving, it is important that considerable care is exercised at the setting-out stage and when jointing. It is often the case that a different type of mortar is required for each material used to form the paving. The following recommendations should be kept to when laying compound paved areas.

(1) Always set out the brick paving first.
(2) Check that the dimensions for slabs do not involve considerable cutting.
(3) The brick paving should always be laid first and jointing should be completed before a start is made on laying the other paving materials.
(4) Where the paving is internal, and longitudinal joints separate the different paving materials, movement joints should be inserted between the different materials.
(5) The completed paved area should be matured before traffic is allowed on the paved surface.

When there is a requirement for brick paving to contain pockets of concrete infill, the paving should be completed and pointed before the *in-situ* concrete infill is placed. During placing of the concrete it is good practice to protect the brick paving around each pocket with lightweight plastic sheeting to prevent staining of the brick paving.

9
QUANTITIES OF MATERIALS

BRICKS AND MORTAR

To calculate the number of bricks and the amount of mortar required for any project, the procedure is quite straightforward.

(1) Calculate the area of brickwork, deducting from this figure the area of any openings. The area is found by multiplying the length by the height and it is important to carry out all calculations in metres. For example, if a window opening is given as 600 mm by 600 mm these figures must be multiplied as 0.6 x 0.6 m.

(2) Multiply the area by the number of bricks per square metre, where

thickness of walling	no. of bricks/m^2
½ brick	60
1 brick	120

For example, in 5 m^2 of half-brick-thick walling there are 5 x 60 = 300 bricks.

(3) Add a percentage (usually 5 per cent) to compensate for any damaged bricks, etc. Perhaps the simplest way to find 5 per cent of any number is first to find 10 per cent by moving the point backwards one place and dividing the figure obtained by 2. For example, if 300 bricks are required, then 10 per cent of this is 30 and 5 per cent is 15. Thus the total number of bricks required is 300 + 15 = 315.

(4) While for accuracy mortar should be ordered in cubic metres per square metre of brickwork, it is sufficiently accurate for the craft student to understand that 1 tonne of mortar is enough to lay 1000 bricks, because this very much simplifies the procedure. Since 1 tonne is 1000 kg, it takes 1000 kg to lay 1000 bricks, which is 1 kg per brick. Therefore it will take 315 kg (0.315 tonnes) to lay 315 bricks, and 1.315 tonnes for 1315 bricks, etc.

Example 9.1

Calculate the number of bricks and the amount of mortar required to complete the area of walling shown in figure 9.1.

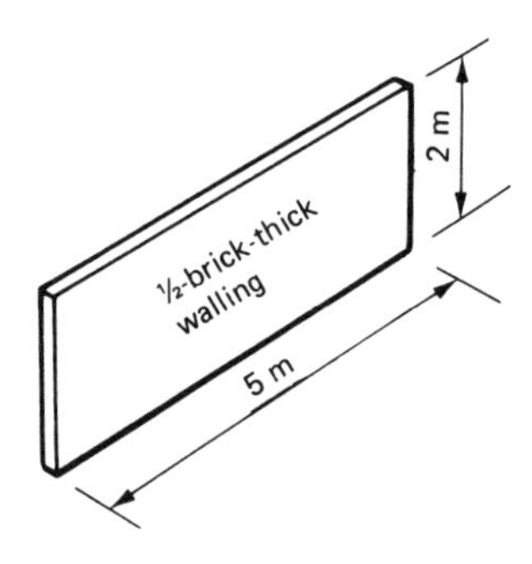

Figure 9.1

Area of walling = length x height

= 5 x 2

= 10 m^2

Number of bricks required = area x no./m^2

= 10 x 60

= 600

Add 5%

10% = 60.0 (move the point back one place)

Therefore

5% = 30 (dividing by 2)

Total number of bricks required = 600 + 30

= 630

and

amount of mortar = 630 kg (0.63 tonnes)

Example 9.2

Calculate the number of bricks and the amount of mortar required to complete the area of walling shown in figure 9.2.

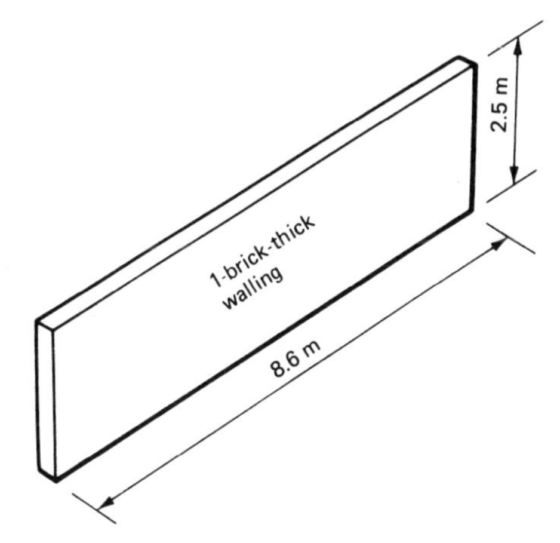

Figure 9.2

Area = length x height

= 8.6 x 2.5

= 21.5 m^2

```
  8.6
  2.5
 1720
  430
 2150
```

Number of bricks = area x number/m^2

= 21.5 x 120

= 2580

```
  21.5
   120
 21500
  4300
2580.0
```

Add 5%

10% = 258

Therefore

5% = 129

Total number of bricks = 2580 + 129

= 2709

and

amount of mortar = 2.709 tonnes (2 tonnes, 709 kg)

Example 9.3

Calculate the number of bricks and the amount of mortar required to complete the walling shown in figure 9.3.

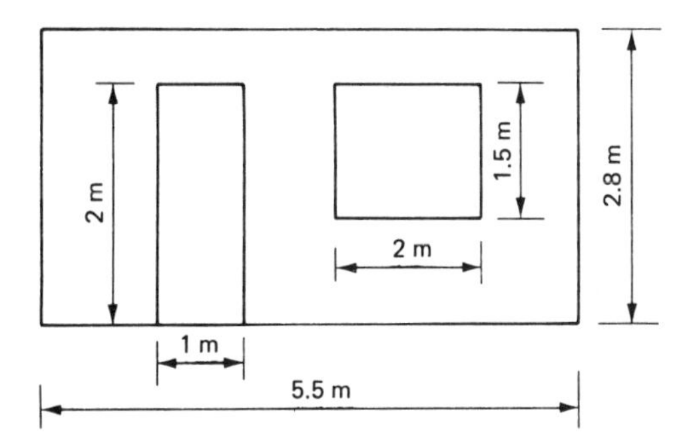

Figure 9.3

Area of brickwork = total area – area of doors and window

total area = length x height

= 5.5 x 2.8

= 15.4 m^2

```
 5.5
 2.8
1100
 440
1540
```

Area of window = length x height

= 2 x 1.5

= 3 m^2

Area of door = length x height

= 1 x 2

= 2 m^2

Total area of openings = 3 + 2

= 5 m^2

Therefore

area of brickwork = 15.4 – 5

= 10.4 m^2

Number of bricks = area x number/m^2

= 10.4 x 60

= 624

Add 5%

10% = 62.4

Therefore

5% = 31.2

= 32 (to the number above)

Therefore

total number of bricks required = 624 + 32

= 656

and

amount of mortar = 656 kg (0.656 tonnes)

Example 9.4

Calculate the number of bricks and the amount of mortar required to build the gable shown in figure 9.4.

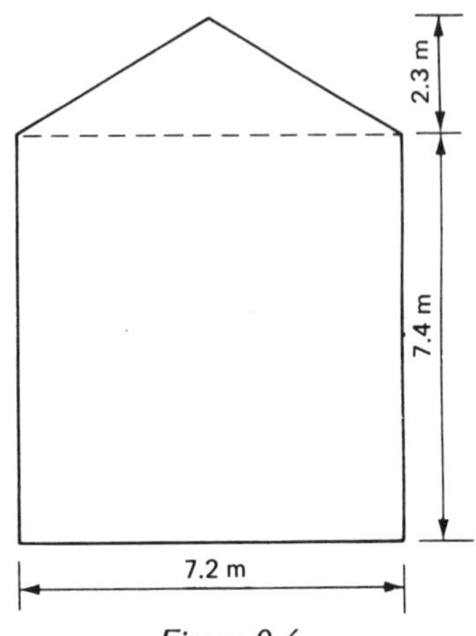

Figure 9.4

Area to eaves = length x height

= 7.2 x 7.4

$= 53.28\ m^2$

```
 7.2
 7.4
5040
 288
5328
```

$$\text{Area of cut-up gable} = \frac{\text{base x height}}{2}$$

$$= \frac{7.2 \times 2.3}{2}$$

= 3.6 x 2.3

$= 8.28\ m^2$

```
3.6
2.3
720
108
828
```

Total area = 53.28 + 8.28

$= 61.56\ m^2$

Number of bricks required = area x number/m^2

= 61.56 x 60

= 3693.6

= 3694 (nearest one above)

```
 61.56
    60
369360
```

Add 5%

10% = 369.4

Therefore

5% = 184.7

= 185 (nearest one above)

Therefore

total number of bricks required = 3694 + 185

= 3879

and

amount of mortar = 3.879 tonnes (3 tonnes, 879 kg)

Example 9.5

Calculate the number of bricks and the amount of mortar required to complete the brickwork shown in figure 9.5.

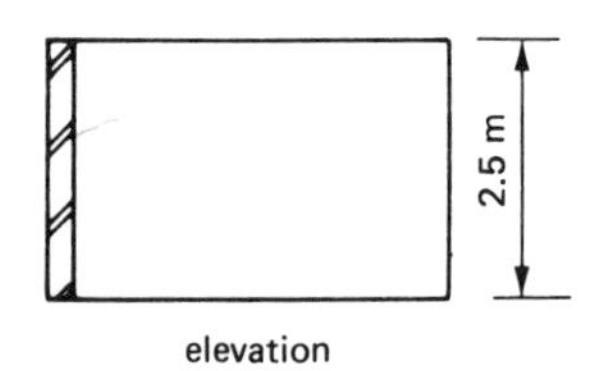

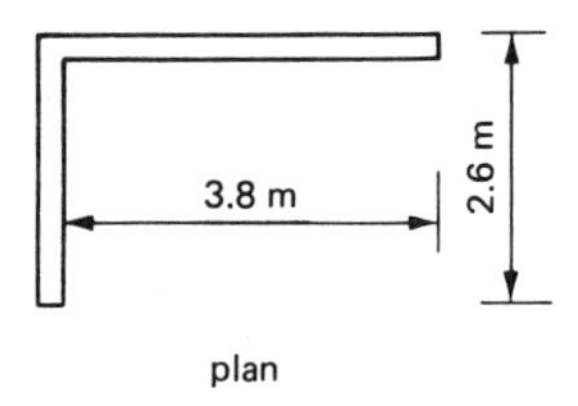

Figure 9.5

Area = length x height (one face) + length x height (other face)

= (3.8 x 2.5) + (2.6 x 2.5)

= 9.5 + 6.5

$= 16\ m^2$

Number of bricks = area x number/m^2

= 16 x 120

= 1920

Add 5%

10% = 192

Therefore

$$5\% = 96$$

$$\text{Total number of bricks} = 1920 + 96$$
$$= 2016$$

and

$$\text{amount of mortar} = 2.016 \text{ tonnes}$$
(2 tonnes, 16 kg)

It should be noted that the area could have been found by multiplying the length of the centre line by the height.

Example 9.6

The inspection chamber shown in figure 9.6 is to be built in engineering bricks class B and is one brick

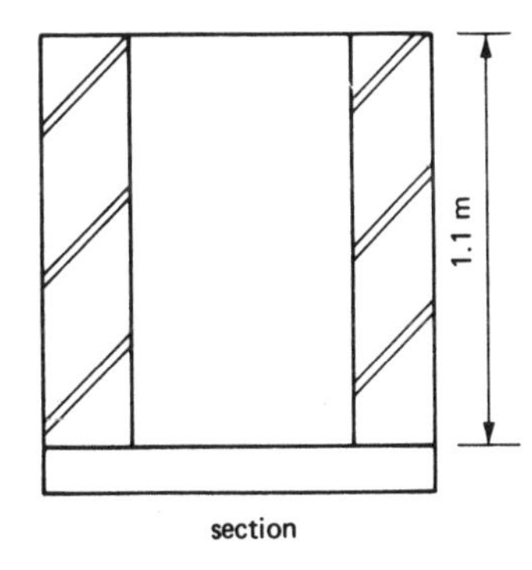

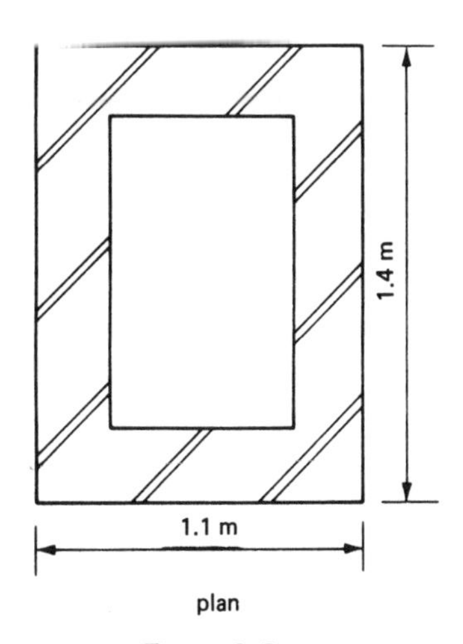

Figure 9.6

thick. Calculate the number of bricks required, including 5 per cent for wastage, and the amount of mortar required.

Length of centre line = (2 x length) + (2 x breadth) − (4 x wall thickness)
(see volume 1, p. 49)

$$= (2 \times 1.4) + (2 \times 1.1) - (4 \times 0.225)$$
$$= 2.8 + 2.2 - 0.9$$
$$= 5 - 0.9$$
$$= 4.1 \text{ m}$$

$$\text{Area} = \text{length of centre line} \times \text{height}$$
$$= 4.1 \times 1.1$$
$$= 4.51 \text{ m}^2$$

```
 4.1
 1.1
 410
  41
 451
```

$$\text{Number of bricks} = \text{area} \times \text{number/m}^2$$
$$= 4.51 \times 120$$
$$= 541.2$$
$$= 542 \text{ (nearest one above)}$$

Add 5%

$$10\% = 54.2$$

Therefore

$$5\% = 27.1$$
$$= 28 \text{ (nearest one above)}$$

Therefore

$$\text{total number of bricks} = 543 + 28$$
$$= 571$$

and

$$\text{amount of mortar} = 571 \text{ kg } (0.571 \text{ tonnes})$$

Example 9.7

Figure 9.7 shows an inspection chamber where the internal dimensions are given. Calculate the quantities of material as for example 9.6.

Length of centre line = (2 x length) + (2 x breadth) + (4 x wall thickness)
(see volume 1, p. 41)

$$= (2 \times 0.95) + (2 \times 0.65) + (4 \times 0.225)$$
$$= 1.9 + 1.3 + 0.9$$
$$= 4.1 \text{ m}$$

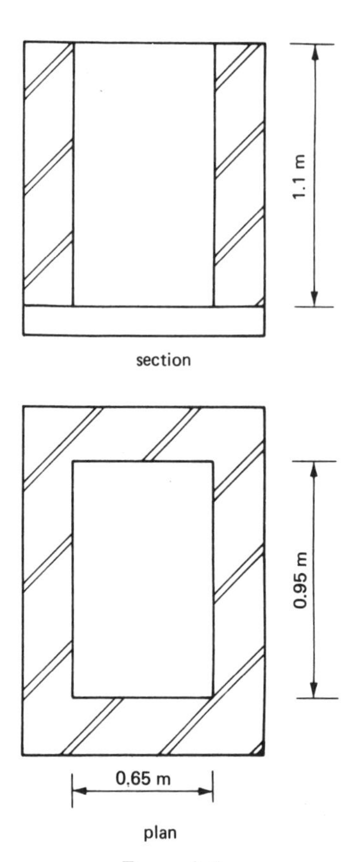

Figure 9.7

It will be noticed by referring to example 9.6 that this is the same length of centre line, and a closer examination will reveal that this is exactly the same inspection chamber, only in this case the internal dimensions have been given. Thus calculations from this point are exactly as shown in example 9.6.

ENGLISH AND FLEMISH BONDS

Where walls are built one brick thick and over with English or Flemish bond, the number of bricks required per square metre, assuming a fair face is necessary on one side only is

English bond	90
Flemish bond	80

Therefore, to calculate the number of facings and commons required for any area of walling, the procedure is as follows.

(1) Calculate the area of brickwork.
(2) Calculate the total number of bricks required (area x number/m^2).
(3) Multiply the area by the number of facings per m^2.
(4) Deduct the number of facings from the total number to obtain the number of commons.

A simple example of each bond is shown below.

Example 9.8

A wall 6 m long and 1.5 m high is to be built in English bond one brick thick. Calculate the number of facings and commons required.

$$\text{Area} = \text{length} \times \text{height} = 6 \times 1.5 = 9\ m^2$$

$$\text{Number of bricks} = \text{area} \times \text{number}/m^2 = 9 \times 120 = 1080$$

$$\text{Number of facings} = 9 \times 90 = 810$$

$$\text{Number of commons} = 1080 - 810 = 270$$

Example 9.9

A wall 5 m long and 2.5 m high is to be built in Flemish bond one brick thick. Calculate the number of facings and commons required.

$$\text{Area} = \text{length} \times \text{height} = 5 \times 2.5 = 12.5\ m^2$$

$$\text{Number of bricks} = \text{area} \times \text{number}/m^2 = 12.5 \times 120 = 1500$$

$$\text{Number of facings} = 12.5 \times 80 = 1000$$

$$\text{Number of commons} = 1500 - 1000 = 500$$

BLOCKS AND MORTAR

There are ten 450 x 100 x 215 mm blocks per square metre, which is a very convenient number for use in calculations. As with brickwork the procedure is: find the area, multiply this by the number per square metre, and add 5 per cent for wastage where an allowance is required.

The amount of mortar required for solid blockwork 100 mm thick is approximately one-third of the quantity required for brickwork of a similar thickness, that is, since one brick takes 1 kg of mortar and a block is equal in area to six bricks, one block takes 2 kg of mortar and 500 blocks take 1 tonne.

Example 9.10

Calculate the number of 215 x 100 x 215 mm blocks and the amount of mortar required to build figure 9.8 including 5 per cent for wastage.

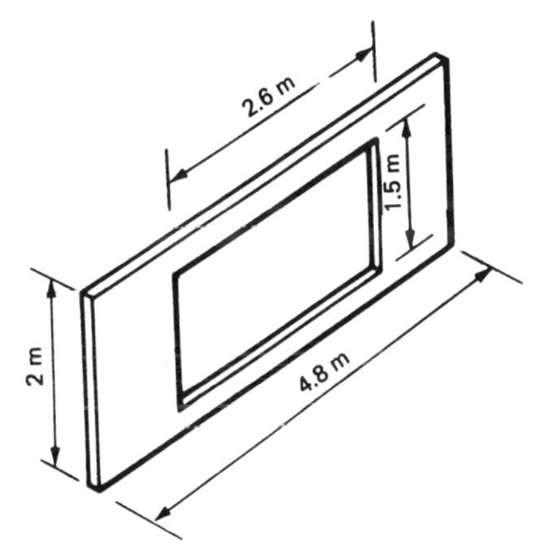

Figure 9.8

Area of blockwork = total area − area of opening

Total area = length x height

= 4.8 x 2

= 9.6 m^2

Area of opening = length x height

= 2.6 x 1.5

= 3.9 m^2

Area of blockwork = 9.6 − 3.9

= 5.7 m^2

Number of blocks = area x number/m^2

= 5.1 x 10

= 51

Add 5%

10% = 5.1

Therefore

5% = 2.55 ≏ 3

Therefore

total number of blocks = 51 + 3

= 54

and amount of mortar = 108 kg (2 kg per block)

Example 9.11

Calculate the number of 100 mm blocks required to build the internal leaf of the gable shown in figure 9.9, including 5 per cent for wastage. Calculate also the amount of mortar.

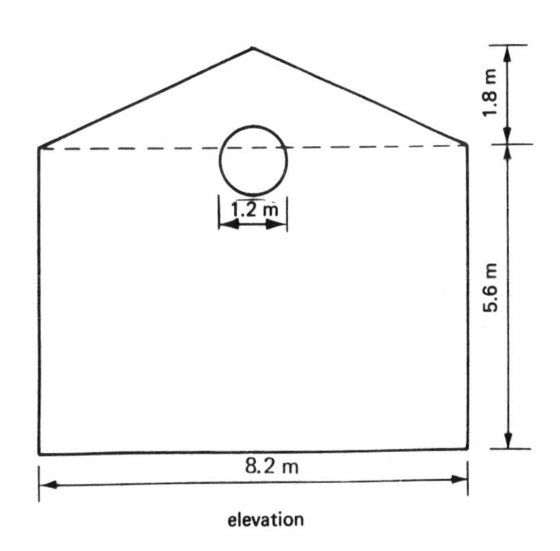

Figure 9.9

Area of blockwork = total area − area of circular opening

Total area = rectangular area + triangular area

$$= \text{length x height} + \frac{\text{base x height}}{2}$$

$$= 8.2 \times 5.6 + \frac{8.2 \times 1.8}{2}$$

= 45.92 + 7.38

= 53.3 m^2

Circular area = πr^2

= 3.142 x 0.6 x 0.6

= 1.1 m^2

Area of blockwork = total area – circular area

= 53.3 – 1.1

= 52.2 m^2

Number of blocks = area x number/m^2

= 52.2 x 10

= 522

Add 5%

10% = 52.2

Therefore

5% = 26.1

= 27 (nearest one above)

Therefore

total number of blocks = 522 + 27

= 549

and amount of mortar = 1.098 tonnes (1098 kg)

VOLUMES OF CONCRETE

The amount of concrete required for any project such as foundations, lintels, paths, slabs, etc. can normally be calculated from the formula

volume = length x breadth x depth

Only for somewhat unusual shapes, for example, circular columns, triangular sections, is a different formula required. It is important to carry out all calculations, except for very small quantities, in cubic metres (m^3) and therefore where the length, breadth or depth is given in millimetres it must be converted to metres before commencement. For example, to multiply 5 m by 100 mm by 50 mm the figures would be 5 x 0.1 x 0.05 m^3.

Example 9.12

A strip foundation for a boundary wall is 12.5 m long, 600 mm wide and 150 mm thick. Calculate the volume.

Volume = length x breadth x depth

= 12.5 x 0.6 x 0.15

= 7.5 x 0.15

= 1.125 m^3

Example 9.13

A concrete path is 6.4 m long, 950 mm wide and 60 mm thick. Calculate the volume.

Volume = length x breadth x depth

= 6.4 x 0.95 x 0.06

= 6.08 x 0.05

= 0.3648 m^3

Example 9.13

A concrete lintel is 1.2 m long, 100 mm wide and 150 mm deep. Calculate the volume.

Volume = length x breadth x depth

= 1.2 x 0.15 x 0.1

= 0.18 x 0.1

= 0.018 m^3

Example 9.14

A constructional hearth has the measurements shown in figure 9.10. Calculate the volume of concrete required.

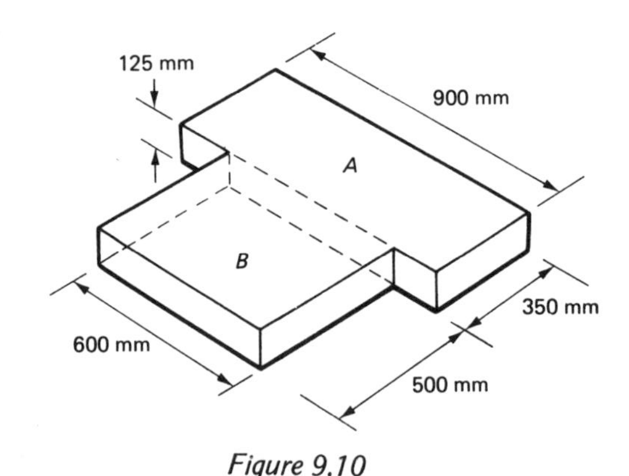

Figure 9.10

Note This must be divided into two parts, as shown by the dotted line. The volume of each part is then calculated separately and the parts are added together to obtain the total volume.

Volume *A* = length x breadth x depth

= 0.9 x 0.35 x 0.125

= 0.315 x 0.125

= 0.039375 m^3

Volume B = length x breadth x depth
= 0.6 x 0.5 x 0.125
= 0.3 x 0.125
= 0.0375 m^3

Total volume = $A + B$
= 0.039375 + 0.0375
= 0.076875 m^3

DRY MATERIAL REQUIREMENTS

Concrete

When concrete is to be mixed on site it is sometimes necessary to determine the amounts of cement, fine and coarse aggregate for ordering purposes. While the absolute volume method is the most accurate, because it involves the use of specific gravities it is dealt with in the advanced volume. The following method gives figures slightly in excess of those required.

The density (mass per cubic metre) of well-compacted concrete is approximately 2400 kg/m^3, and therefore if the mix is to be 1:2:4 the density should be divided by 7 since there are seven parts (1 + 2 + 4) to give the amount of cement required. This is doubled for the amount of sand and multiplied by 4 for the coarse aggregate. For example

cement = 2400 ÷ 7 = 342 kg (just under 7 bags)
sand = 342 x 2 = 684 kg
stone = 342 x 4 = 1368 kg

Similarly, if the mix is to be 1:3:6 and the density is 2400 kg/m^3, the density is divided by 10 since there are 10 parts (1 + 3 + 6) to give the amount of cement. This is multiplied by 3 to obtain the amount of sand and by 6 to obtain the stone. For example

cement = 2400 ÷ 10 = 240 kg (just under 5 bags)
sand = 240 x 3 = 720 kg
stone = 240 x 6 = 1440 kg

Mixing Shrinkage

Many mistakes are made in examination questions by students using the wrong method to arrive at the answer. Put as simply as possible:

Example 1

If dry materials to make 1 m^3 of concrete are needed for a job and the mixing shrinkage is 50 per cent, it is *wrong* to simply add 50 per cent, because this too will suffer from mixing shrinkage. For example

1 m^3 + 50 per cent = 1.5 m^3
BUT shrinkage is 50 per cent = 0.75 m^3
which produces only 0.75 m^3
WRONG! (we needed 1 m^3)

The correct method is to multiply by $\frac{100}{50}$ as follows:

$1 \text{ m}^3 \times \frac{100}{50}$ = 2 m^3
Shrinkage is 50 per cent = 1 m^3
this produces 1 m^3
(which is what was required)

Example 2

How many cubic metres of dry materials are required to make 6 m^3 of concrete if the mixing shrinkage is 25 per cent.

Here again if we simply add 25 per cent to 6 m^3
this gives us 7.5 m^3
but shrinkage is 25 per cent = 1.875 m^3
which produces = 5.625 m^3
WRONG! (we needed 6 m^3)

Again the correct method is to multiply by $\frac{100}{75}$ as follows:

$6 \text{ m}^3 \times \frac{100}{75}$ = 8 m^3
and shrinkage is 25 per cent = 2 m^3
giving 6 m^3
(which is what was required)

Therefore, given a shrinkage rate of:

(a) 10 per cent, multiply the required volume by $\frac{100}{90}$

(b) 20 per cent, multiply the required volume by $\frac{100}{80}$

(c) 30 per cent, multiply the required volume by $\frac{100}{70}$

(d) 35 per cent, multiply the required volume by $\frac{100}{65}$

etc.

Mortar

The density of mortar is approximately 2300 kg/m^3 and different mixes are used for different situations. For example, a 1:3 cement/sand is normally used for engineering bricks, whereas 1:6 cement/sand plus plasticiser may be specified for facing brickwork. For aerated concrete blocks used for the inner leaf of external cavity walls 1:2:8 is quite strong enough. To calculate the requirements for any mix the density should be divided by the number of parts to obtain the cement content and this figure is then multiplied by the number of parts of sand to obtain the sand content.

Example 9.15

Requirements for a 1:3 mix are

$$\text{cement} = \frac{2300}{4} = 575 \text{ kg } (11\tfrac{1}{2} \text{ bags})$$

$$\text{sand} = 575 \times 3 = 1725 \text{ kg } (1.725 \text{ tonnes})$$

Example 9.16

For a 1:6 mix

$$\text{cement} = \frac{2300}{7} = 328\tfrac{1}{2} \text{ kg (about } 6\tfrac{1}{2} \text{ bags)}$$

$$\text{sand} = 328\tfrac{1}{2} \times 6 = 1971 \text{ kg } (1.971 \text{ tonnes})$$

Example 9.17

For a 1:1:8 mix

$$\text{cement} = \frac{2300}{10} = 230 \text{ kg (just over } 4\tfrac{1}{2} \text{ bags)}$$

$$\text{lime} = \frac{2300}{10} = 230 \text{ kg}$$ (just over 9 bags because a bag of lime contains 25 kg)

$$\text{sand} = 230 \times 8 = 1840 \text{ kg } (1.84 \text{ tonnes})$$

NUMBER OF FLOOR TILES OR PAVING SLABS

To calculate the number of floor tiles or paving slabs required to cover any given area, the procedure is as follows

(1) find the total area to be tiled in square metres
(2) find the area of each tile in square metres
(3) divide the total area to be tiled by the area of each tile
(4) add a percentage for wastage where an allowance is required.

Example 9.18

A room measures as shown in figure 9.11. Calculate the number of 300 x 300 mm tiles required to cover this area.

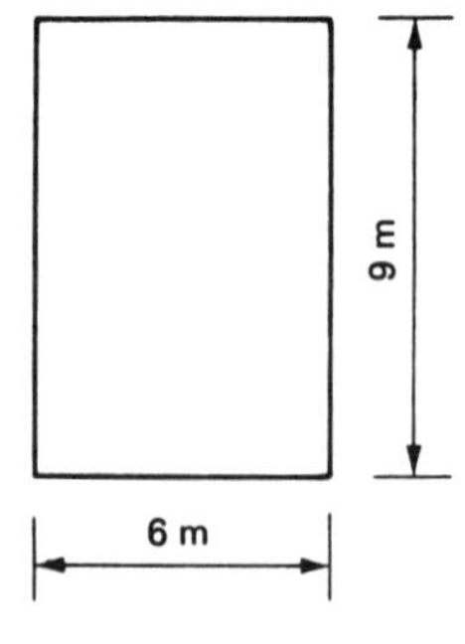

Figure 9.11

$$\text{Area} = \text{length} \times \text{breadth} = 9 \times 6 = 54 \text{ m}^2$$

$$\text{Area of each tile} = \text{length} \times \text{breadth} = 0.3 \times 0.3 = 0.09 \text{ m}^2$$

$$\text{Number of tiles required} = \frac{\text{total area}}{\text{area of each tile}} = \frac{54}{0.09} = 600$$

$$9\overline{)5400} = 600; \quad 54, \; 00$$

Example 9.19

Ignoring wastage, find the number of 500 x 500 mm paving slabs required to pave the circular area shown in figure 9.12.

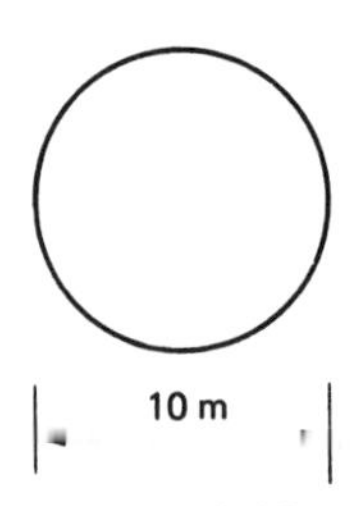

Figure 9.12

$$\text{Area} = \pi r^2 = 3.142 \times 5 \times 5 = 78.55 \text{ m}^2$$

$$\text{Area of each slab} = \text{length} \times \text{breadth} = 0.5 \times 0.5 = 0.25 \text{ m}^2$$

$$\text{Total number of slabs} = \frac{\text{total area}}{\text{area of each slab}}$$

$$= \frac{78.55}{0.25}$$

$$= 314.2$$

$$= 315$$

```
    314.2
25)7855.0
   75
    35
    25
    105
    100
      50
      50
```

Example 9.20

Adding 5 per cent for wastage, calculate the number of 900 x 600 mm paving slabs required to pave the area shown in figure 9.13.

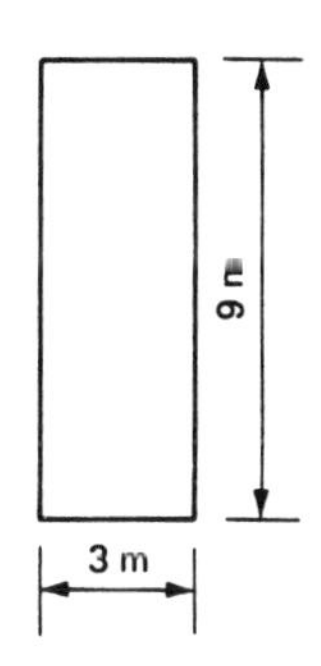

Figure 9.13

$$\text{Area} = \text{length} \times \text{breadth}$$

$$= 9 \times 3$$

$$= 27 \text{ m}^2$$

$$\text{Area of each slab} = \text{length} \times \text{breadth}$$

$$= 0.9 \times 0.6$$

$$= 0.54 \text{ m}^2$$

$$\text{Number of slabs required} = \frac{\text{total area}}{\text{area of each slab}}$$

$$= \frac{27}{0.54}$$

$$= 50$$

Add 5%

10% = 5

5% = 2½

Total number required = 53

Example 9.21

Ignoring wastage, calculate the number of tiles 200 mm square required to complete the floor area shown in figure 9.14.

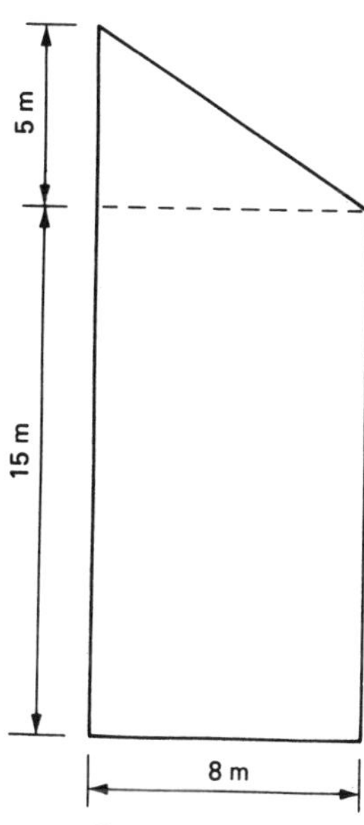

Figure 9.14

$$\text{Area of rectangle} = \text{length} \times \text{breadth}$$

$$= 15 \times 8$$

$$= 120 \text{ m}^2$$

$$\text{Area of triangle} = \frac{\text{base} \times \text{height}}{2}$$

$$= \frac{8 \times 5}{2}$$

$$= 20 \text{ m}^2$$

$$\text{Total area} = 120 + 20$$

$$= 140 \text{ m}^2$$

$$\text{Area of each tile} = \text{length} \times \text{breadth}$$

$$= 0.2 \times 0.2$$

$$= 0.04 \text{ m}^2$$

$$\text{Number of tiles required} = \frac{\text{total area}}{\text{area of each tile}}$$

$$= \frac{140}{0.04}$$

$$= 3500$$

```
   35
4)14000
  12
   20
   20
```

Example 9.22

Figure 9.15 shows the plan of a large kitchen floor which is to be tiled with 150 x 150 mm tiles. The area shown shaded is not to be tiled. Calculate the total number of tiles required, making no allowance for wastage.

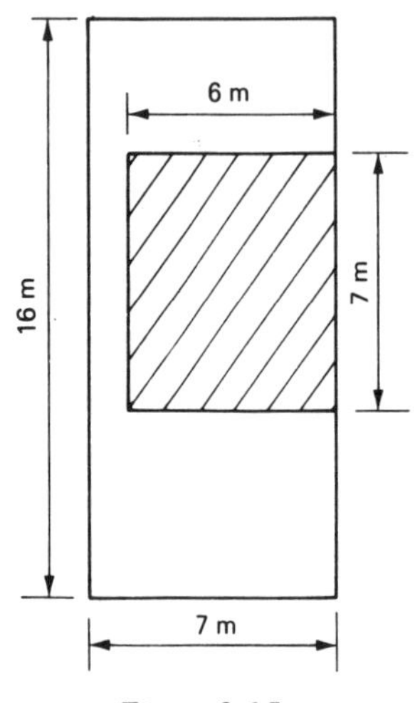

Figure 9.15

Overall area = length x breadth
= 16 x 7
= 112 m^2

Shaded area = length x breadth
= 7 x 6
= 42 m^2

Total area to be tiled = overall area – shaded area
= 112 – 42
= 70 m^2

Area of each tile = length x breadth
= 0.15 x 0.15
= 0.0225 m^2

Number of tiles required $= \dfrac{\text{tiled area}}{\text{area of each tile}}$

$= \dfrac{70}{0.0225}$

= 3111

No.	Log
70	1.8451
0.0225	2.3522
3111	3.4929

Areas of Irregular Figures

The approximate area of an irregular figure may be determined by a number of methods, including the mid-ordinate rule, the squared paper method, the use of measuring instruments, and Simpson's rule. It is usually considered sufficient for craft students to have an understanding of the mid-ordinate rule which is explained as follows.

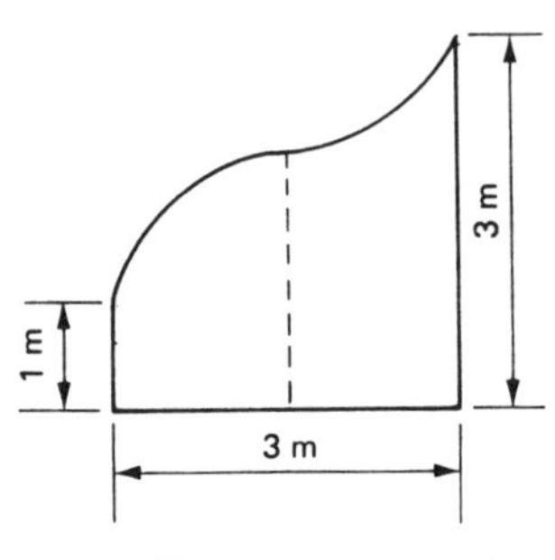

Figure 9.16

Consider figure 9.16. The approximate area of this figure may be calculated by multiplying the width by the average height or length, shown by the broken line, which is known as the mid-ordinate. That is

area = width x mid-ordinate
= 3 x 2
= 6 m^2

Figure 9.17 shows a larger, more complicated area, and for accuracy this has been divided into a convenient number of strips of equal width. The mid-ordinates are again shown by broken lines.

Area of first strip = width x mid-ordinate
= 3 x 2.5 = 7.5 m^2

Area of secnd strip = width x mid-ordinate
= 3 x 3.5 = 10.5 m^2

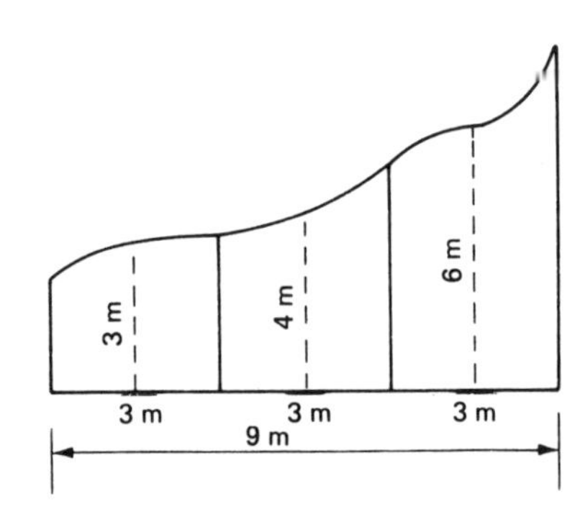

Figure 9.17

Area of third strip = width x mid-ordinate
$= 3 \times 5 = 15 \text{ m}^2$

Total area = $7.5 + 10.5 + 15 = 33 \text{ m}^2$

A simpler way is to say

Area = width of strip x sum of the mid-ordinates

$= 3 \times (2.5 + 3.5 + 5)$

$= 3 \times 11$

$= 33 \text{ m}^2$

Figure 9.18 represents an area of land 12 m wide. To find the approximate area this must be divided into a number of strips of equal width and the mid-ordinate of each strip drawn and scaled off as shown. The length could have been divided into three 4 m wide strips, four 3 m wide strips, twelve 1 m wide strips or, as shown, six 2 m strips; it should be obvious that the greater the number of strips, the more accurate will be the result.

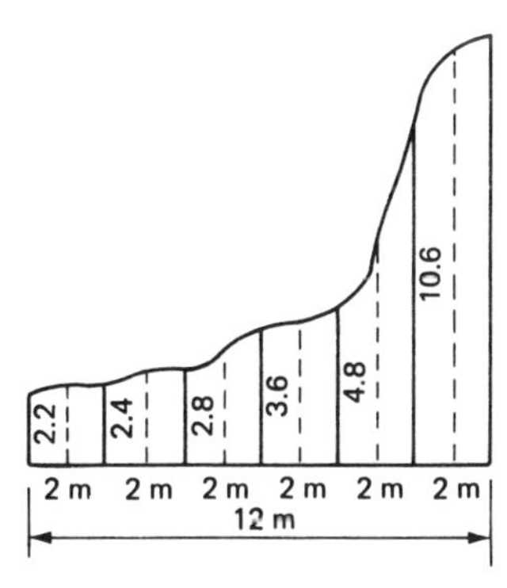

Figure 9.18

To calculate the area using the formula given

Area = width of strip x sum of the mid-ordinates

$= 2 \times (2.2 + 2.4 + 2.8 + 3.6 + 4.8 + 10.6)$

$= 2 \times 26.4$

$= 52.8 \text{ m}^2$

Figure 9.19 shows an area of land 16 m wide which has been divided into four strips, each 4 m in width. The mid-ordinates are drawn and accurately scaled off as before.

Area = width of strip x sum of the mid-ordinates

$= 4 \times (6.4 + 10.4 + 12.8 + 14.2)$

$= 4 \times 43.8$

$= 175.2 \text{ m}^2$

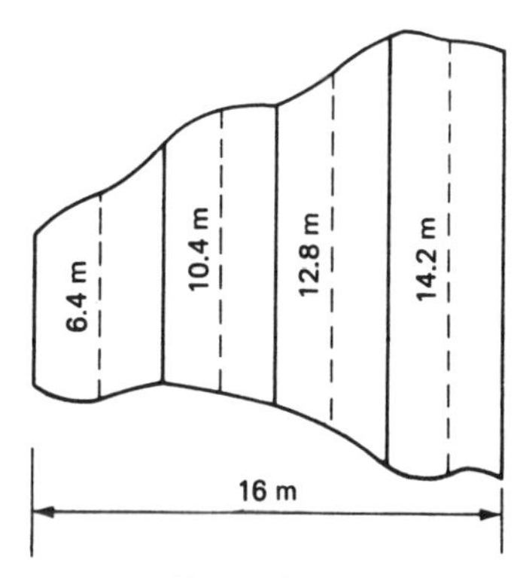

Figure 9.19

Note Figure 9.20 shows why the mid-ordinate rule is fairly accurate; the formula directs that the strips are placed end to end and their overall length is then multiplied by the width of the strip.

One final example is as follows. The irregular figure shown in figure 9.21 is to be covered with a layer of concrete 150 mm in depth. Calculate the volume of concrete required.

Area = width of strip x sum of the mid-ordinates

$= 4 \times (5 + 6 + 8 + 10 + 12 + 11 + 8)$
(mid-ordinates averaged)

Area = 4×60

$= 240 \text{ m}^2$

Volume = area x thickness

$= 240 \times 0.15$

$= 36 \text{ m}^3$

Note If the ordinates are all given, the mid-ordinates may be averaged (figure 9.21). If not they must be scaled off (figure 9.19).

BRICKS FOR PAVING

When brick-on-edge work is being carried out the number of bricks per square metre is as for face brickwork, that is, 60; but for brick flat paving there are 45 per square metre.

Example 9.23

Calculate the number of paviors required to pave a 8.6 m x 7.5 m courtyard with brick flat paving.

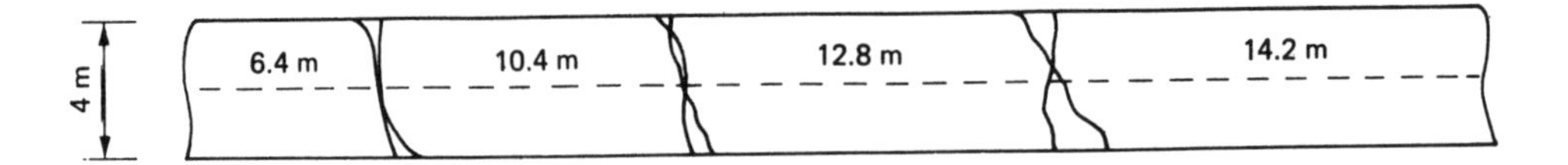

Figure 9.20 Figure 9.19 drawn by formula

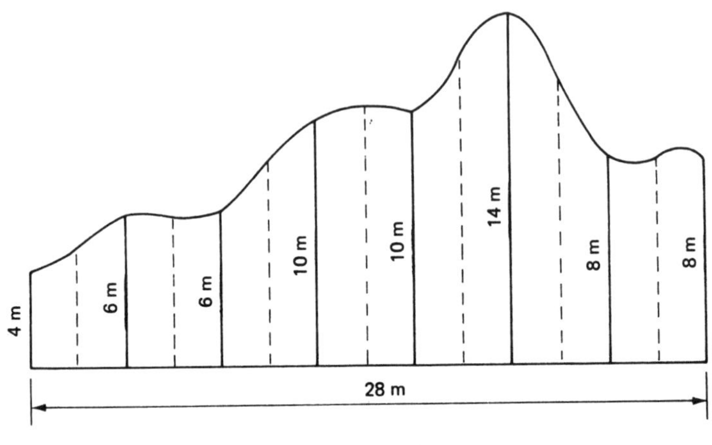

Figure 9.21

Area = length x breadth

= 8.6 x 7.5

= 64.5 m^2

Number of bricks = area x number/m^2

= 64.5 x 45

= 2902.5

= 2903

SOME USEFUL CONVERSION FACTORS

To convert to metric, multiply by the factor given.
To convert from metric, divide by the same factor.

	factor
1. *Length*	
inches to millimetres	25.4
feet to metres	0.3048
yards to metres	0.9144
miles to kilometres	1.6093
2. *Area*	
square feet to square metres	0.0929
square yards to square metres	0.8361
acres to square metres	4046.86
acres to hectares	0.4047
square miltes to square kilometres	2.59
square miles to hectares	258.999

	factor
3. *Volume*	
cubic inches to cubic millimetres	1639.0
cubic feet to cubic metres	0.0283
cubic yards to cubic metres	0.7646
4. *Mass*	
ounces to grams	28.349
pounds to grams	453.6
pounds to kilograms	0.4536
hundredweights to kilograms	50.8023
tons to tonnes	1.016
tons to kilograms	1016.05
5. *Capacity*	
pints to litres	0.568
quarts to litres	1.137
gallons to litres	4.546

MULTIPLE CHOICE QUESTIONS

Select your options from the questions below, underline your selection, for example (b), and check your answers with those on page 130.

1. The number of site operatives who have lost their lives during the last eight years because of accidents is just over:
   (a) 100
   (b) 500
   (c) 1000
   (d) 2000

2. When using the third order of levers, the effort is applied:
   (a) upwards at one end
   (b) upwards between load and fulcrum
   (c) downwards at one end
   (d) downwards between load and fulcrum

3. Using a pulley, an effort of 400 N will raise a load creating a force of 700 N. The mechanical advantage of the pulley is:
   (a) 0.571
   (b) 2.8
   (c) 28
   (d) 1.75

4. Employees must take reasonable care of the health and safety both of themselves and all other persons who may be affected by their acts or omissions. This is stated in the:
   (a) Construction Regulations
   (b) Health and Safety at Work Act
   (c) Building Regulations
   (d) Building Safety Manual

5. A fanguard on a scaffold should have a minimum slope of:
   (a) 1 in 6
   (b) 1 in 9
   (c) 1 in 12
   (d) 1 in 18

6. When industrial premises require brick paved areas, the type of brick most suitable would be:
   (a) commons
   (b) pressed facings
   (c) sand limes
   (d) engineerings

7. A ½ brick thick wall is 7.000 x 2.00 m. The number of bricks required will be:
   (a) 960
   (b) 840
   (c) 920
   (d) 880

8. The recess in an indented quoin is usually kept to a maximum of:
   (a) 28 mm
   (b) 32 mm
   (c) 38 mm
   (d) 56 mm

9. When single herringbone bond is used for rectangular panels, the cut bricks on each side will be:
   (a) similar
   (b) differing lengths
   (c) opposite
   (d) alternate lengths

10. Clay drains which pass under buildings should be surrounded with:
    (a) 200 mm of broken bricks
    (b) 100 mm of pea gravel
    (c) 150 mm of concrete
    (d) 125 mm of sand

11. The number of facing bricks needed per square metre in English, stretcher and Flemish bonds, respectively, are
    (a) 120, 80, 90
    (b) 80, 60, 90
    (c) 80, 90, 60
    (d) 90, 60, 80

12. The typical sectional size of a poling board used for timbering to trenches is:
    (a) 50 x 25 mm
    (b) 150 x 100 mm
    (c) 220 x 37 mm
    (d) 220 x 100 mm

13. According to the Construction Regulations, the minimum width of a scaffold for men and

materials is:
(a) 2 planks wide
(b) 3 planks wide
(c) 4 planks wide
(d) 5 planks wide

14. A bag of lime has a mass of 25 kg. Because of the force of gravity this creates a force of approximately:
(a) 2.5 N
(b) 25 N
(c) 250 N
(d) 250 kN

15. The property of a material to allow water to pass through it is called:
(a) capillarity
(b) permeability
(c) porosity
(d) absorption

16. If 150 kg of cement is used in a concrete mix and the water/cement ratio is 0.6, the number of litres of water needed is:
(a) 60
(b) 90
(c) 100
(d) 150

17. A well-graded aggregate may be described as:
(a) limestone chippings
(b) particles of a similar size
(c) 1:3:6 cement, sand and stone
(d) from the coarsest specified down to dust

18. When constructing tumbling-in, the ratio of tumbled courses to horizontal courses should be:
(a) 3:2 or 3:4
(b) 4:2 or 3:2
(c) 6:4 or 4:2
(d) 3:4 or 6:4

19. Bricks used for corbelling should be:
(a) soaked before use
(b) sprayed and damped
(c) dry and clean
(d) only slightly damp

20. Quetta bond contains internal pockets within the wall thickness which measure:
(a) 56 x 102 mm
(b) 102 x 102 mm
(c) 168 x 102 mm
(d) 56 x 56 mm

21. The following items can be found in a first aid box:
(a) triangular bandage, scissors, sterile dressings, safety pins
(b) sterile eye pads, safety pins, triangular bandage, adhesive dressings
(c) triangular bandage, guidance leaflet, aspirins, sterile eye pads
(d) safety pins, sterile eye bath, triangular bandage, adhesive dresssings

22. Inlets to drains should have a minimum water seal of:
(a) 32 mm
(b) 100 mm
(c) 75 mm
(d) 50 mm

23. The ball test on drain pipes is used as a check for:
(a) well worn pipes
(b) lipping of joints
(c) any leakage in pipes
(d) correct amount of fall

24. The scaffold fitting used to connect a cross brace to a standard is a:
(a) double coupler
(b) universal coupler
(c) swivel coupler
(d) sleeve coupler

25. Sole plates are used in scaffolding:
(a) when ground bearing strength is suspect
(b) if there are no window openings
(c) to accommodate varying standard lengths
(d) where independent scaffolds are required

26. Before a hoarding can be erected in a public thoroughfare, the contractor must obtain a licence from the:
(a) The Health and Safety inspector
(b) local authority
(c) building control office
(d) licensing authority

27. The simplest herringbone panel to set out and build is:
(a) single
(b) double
(c) feather
(d) diagonal

28. When terminating an attached pier by tumbling-in, if the number of horizontal courses and the number of inclined courses are both even, the result will be:
(a) broken bond below the tumbling courses
(c) no problems with the bonding
(c) straight joints occurring above the tumbling courses
(d) broken bond above the tumbling courses

29. How many cubic metres of dry materials are needed to make 4 m^3 of concrete if mixing shrinkage is 20 per cent?
(a) 4.4
(b) 4.8
(c) 5.0
(d) 6.0

30. To provide intermediate support for ground floor joists, the following should be provided:
(a) fender walls
(b) purlin walls
(c) loadbearing walls
(d) sleeper walls

31. A chimney stack measures 900 x 450 on plan. Its maximum height including the terminal would be:
(a) 900 mm
(b) 1.8 m
(c) 2.0 m
(d) 2.025 m

32. The test on fine aggregate which involves the use of a salt solution is the test for:
(a) organic impurities
(b) efflorescence
(c) silt content
(d) mixing shrinkage

33. The minimum projection for a ladder above the top of a working platform is:
(a) 1.2 m
(b) 1.0 m
(c) 900 mm
(d) 600 mm

34. Leaning chimney stacks often indicate:
(a) sulphate attack
(b) poor workmanship
(c) defective pointing
(d) roof subsidence

35. Cat or crawling ladders are used:
(a) within buildings
(b) in trench excavation
(c) between scaffolds
(d) for roof work

36. A rubber-headed mawl is used to:
(a) compact hardcore fill
(b) fix paving slabs
(c) lay drain pipes
(d) level brick paving

37. The vertical height of a ladder is 5.0 m, so the amount of horizontal distance to the foot of the ladder will be:
(a) 1.250 m
(b) 1.500 m
(c) 1.750 m
(d) 1.150 m

38. A tapered straightedge is used in conjunction with:
(a) sight rails
(b) foundation brickwork
(c) drain pipes
(d) stepped foundations

39. The water-test requires a head of water at the highest point not exceeding:
(a) 1.000 m
(b) 1.500 m
(c) 1.400 m
(d) 1.200 m

40. In drainage systems rigid joints are formed by using:
(a) plastic cement and fibre glass rope
(b) fibre glass rope and cement mortar
(c) tarred gaskin and plastic cement
(d) cement mortar and tarred gaskin

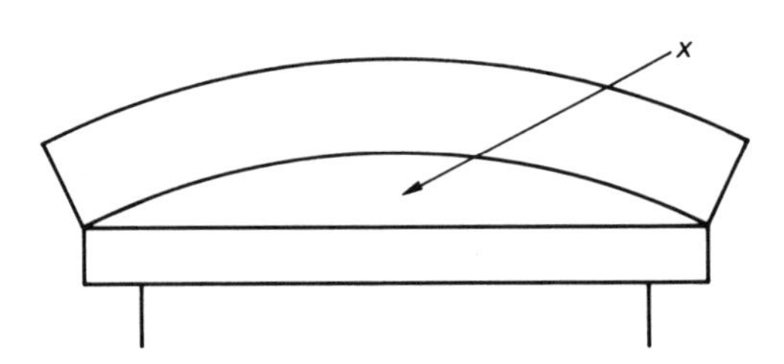

41. The figure illustrates a relieving arch over a lintel.

The part marked *x* is the:
(a) turning piece
(b) core
(c) centre
(d) intrados

42. A double coupler is used on a scaffold:
(a) to fasten transoms to ledgers
(b) in all positions where strength is required
(c) for all longitudinal bracing
(d) for all transverse bracing

43. If a working drawing is made to a scale of 1:50, a length of 7.5 m would be shown on the drawing by a length of:
(a) 150 mm
(b) 250 mm
(c) 75 mm
(d) 15 mm

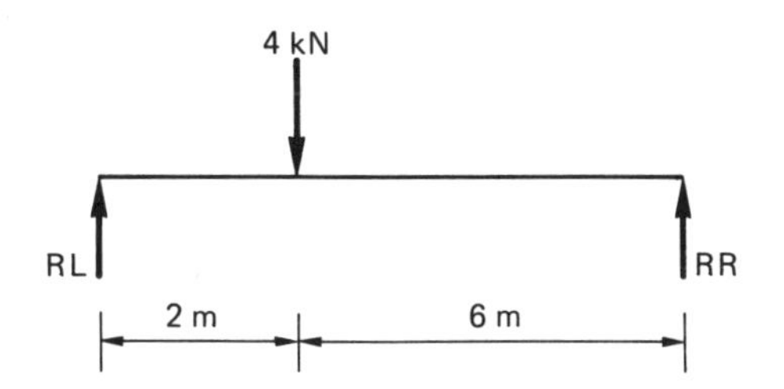

44. What will the reactions be at RL and RR in the problem shown in the figure?

(a) 3 kN and 1 kN
(b) 4 kN and 2 kN
(c) 2 kN and 4 kN
(d) 3 kN and 1 kN

45. Densities of building materials are usually stated in:
(a) N/mm^3
(b) kg/m^3
(c) kN/m^2
(d) kg/m^2

46. Step irons used in manholes should be spaced at a maximum vertical distance of:
(a) 225 mm
(b) 350 mm
(c) 300 mm
(d) 200 mm

47. The internal surfaces of formwork to concrete lintels should be coated with mould oil in order to:
(a) apply a particular finish
(b) prevent leakage of cement grout
(c) assist the striking process
(d) produce a waterproof surface

48. The drainage fitting which is used at the bottom of a soil pipe is a:
(a) rest bend
(b) gulley trap
(c) rain-water shoe
(d) back-inlet gulley

49. A saddle fitting is used to:
(a) reduce the rate of flow
(b) connect drains to sewers
(c) form outlets from manholes
(d) inspect long lengths of pipes

50. Guard rails are provided when operatives are liable to fall more than:
(a) 1.5 m
(b) 2.0 m
(c) 1.8 m
(d) 2.0 m

51. The horizontal distance between standards is termed:
(a) span
(b) spacing
(c) lift
(d) bay

52. The maximum projection of a scaffold board over its end support is equal to its thickness multipied by:
(a) 2
(b) 3
(c) 4
(d) 6

53. Sand-lime bricks are subjected to steam-curing in order to:
(a) harden them
(b) dry them before firing
(c) lower the moisture content
(d) ensure an even colour

54. The slump test is commonly used to test concrete for:
(a) final strength
(b) workability
(c) water content
(d) setting and hardening

55. The elevation shown in the figure is an example of:

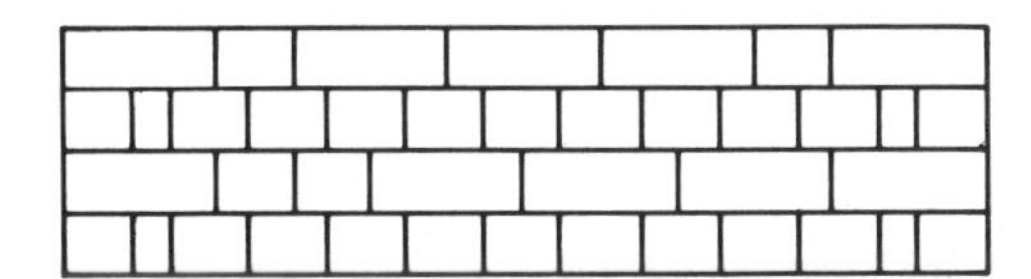

(a) English bond
(b) reverse bond
(c) English cross bond
(d) Dutch bond

56. The underside of a segmental arch is called the:
(a) skewback
(b) intrados
(c) haunch
(d) soffit

57. The best way to reduce heat losses through the external walls of a domestic building is to:
(a) build the inner leaf 1 brick thick in engineerings
(b) have an unventilated cavity between the two leaves
(c) use 100 mm aerated concrete blocks for the inner leaf
(d) fill the cavity with fibre glass

58. Drainpipes are no longer made or used in the UK in the following material:
(a) uPVC
(b) concrete
(c) clay
(d) pitch fibre

59. Flue liners for solid fuel appliances should be bedded in:
(a) the same mortar used for the brickwork of the stack
(b) high alumina cement mortar
(c) sulphate resisting cement mortar
(d) lime sand mortar

60. Flexible jointed pipes should be bedded on:
(a) pea gravel
(b) clinker ash
(c) river sand
(d) crushed bricks

61. The figure shows a:

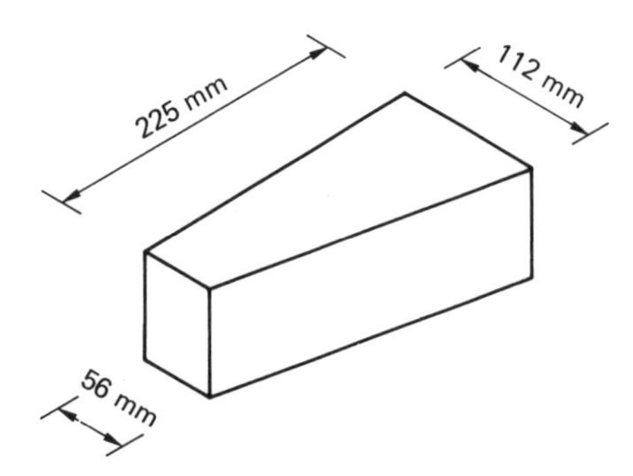

(a) king closer
(b) bevelled closer
(c) mitred bat
(d) cant brick

62. The figure shows a plan course of a 1½ brick square detached pier in:

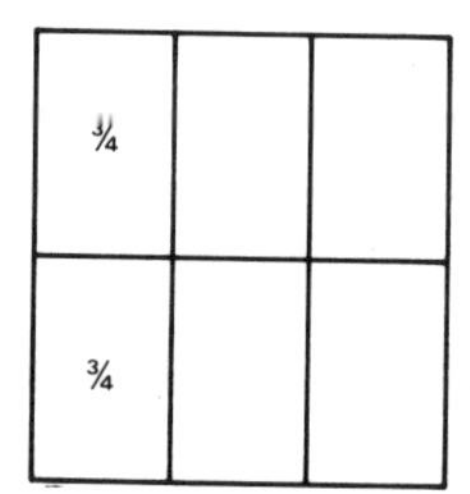

(a) English bond
(b) stretcher bond
(c) Flemish bond
(d) Flemish garden wall bond

63. The figure shows a sectional elevation through a concrete strip foundation. In order to comply with Building Regulations, the dimension marked *x* should be:

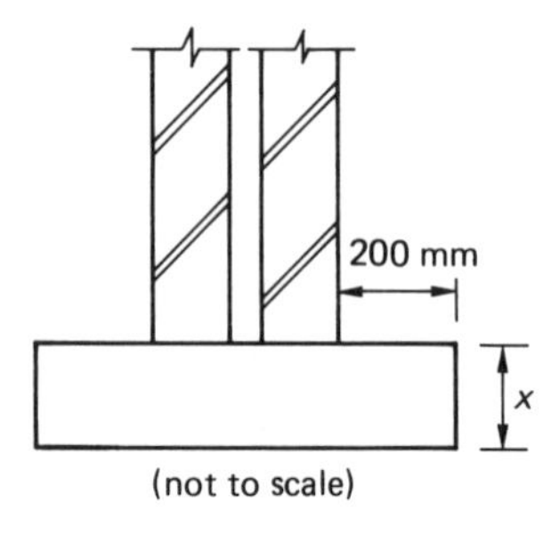

(a) 100 mm
(b) 150 mm
(c) 200 mm
(d) 300 mm

64. The figure shows the first three bricks needed for the commencement of a decorative panel in:

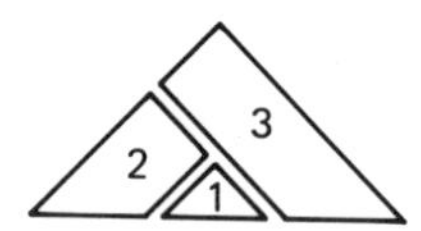

(a) diagonal basket-weave
(b) single herringbone
(c) diagonal herringbone
(d) basket-weave

65. Vertical reinforcement within brick walls is used to:
(a) combat compressive loading
(b) resist lateral pressure
(c) prevent moisture penetration
(d) strengthen weak mortar

66. When walls one brick in thickness have to be vertically reinforced, the bond most suitable would be:
(a) header bond
(b) English bond
(c) modern face bond
(d) rat trap bond

67. The minimum going for a concrete tread to domestic stairs is:
(a) 200 mm
(b) 300 mm
(c) 350 mm
(d) 220 mm

68. Brick steps should have a rise not exceeding:
(a) 200 mm
(b) 350 mm

(c) 300 mm
(d) 220 mm

69. The minimum height for a handrail to a staircase, above the pitch line, is:
(a) 780 mm
(b) 840 mm
(c) 740 mm
(d) 880 mm

70. Drawings (*a*) and (*b*) in the figure show the foot of a raking strut, resting on a concrete base. The strut is being tightened up by means of a lever. The order of levers shown in (*a*) and (*b*) respectively is the:

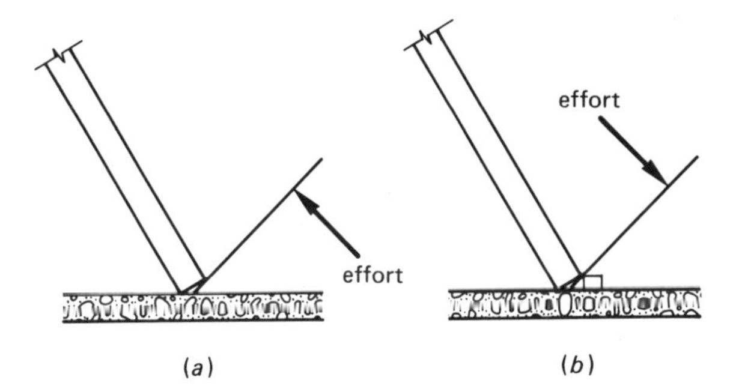

(a) first and second
(b) second and third
(c) first and third
(d) second and first

71. One reason why diagonal herringbone is the easiest herringbone panel to set out and build is because:
(a) all the work is at 45°
(b) no diagonal cutting is necessary
(c) all the cut bricks are the same size
(d) all bricks to be cut can be marked from the same bevel

72. Given that the mechanical advantage of a certain pulley is 3, an effort of 150 N will raise a load creating a force of:
(a) 50 N
(b) 150 N
(c) 300 N
(d) 450 N

73. When carrying out the slump test on a sample of concrete, the number of layers in which the cone is filled and the number of times each layer is rodded is:
(a) 3 and 25
(b) 4 and 25
(c) 3 and 35
(d) 4 and 35

74. A dentil course should not project more than:
(a) 75 mm
(b) 28 mm
(c) 56 mm
(d) 12 mm

75. A string course is built:
(a) around the face of a building
(b) over window openings
(c) under certain types of arches
(d) below window sills

76. Square diagonal basketweave panels are set out:
(a) below the centre line
(b) from the sides of the panel
(c) at the base of the panel
(d) from the centre

77. Plinth courses are normally used to:
(a) form a decorative effect
(b) increase the wall thickness
(c) provide an alternative bond
(d) reduce the wall thickness

78. The equipment required to carry out tumbling-in work is:
(a) bevel, square, lines, templet
(b) square, templet, gun, bevel
(c) lines, square, templet, gun
(d) gun, lines, bevel, square

79. Corbel courses are used to:
(a) form string courses
(b) increase wall thickness
(c) increase stability
(d) terminate piers

80. The height, base diameter and top diameter of the slump cone are respectively:
(a) 300, 200, 100 mm
(b) 300, 150, 100 mm
(c) 200, 150, 100 mm
(d) 300, 200, 150 mm

81. The Factory Inspector must be notified:
(a) if materials delivered to site do not comply with a BS
(b) if a hoarding is to be erected in a public thoroughfare
(c) where an injury involves absence from work for 3 days or more
(d) before pouring concrete in an excavation for a foundation

82. A header is placed adjacent to the quoin three-quarter on every other course of stretchers. This bond is:
(a) English garden wall
(b) English cross

(c) Dutch
(d) monk chevron

83. A label course is used:
(a) over the top of an arch
(b) to tie separate rings of an arch together
(c) as a decorative feature at storey height
(d) below a relieving arch on a concrete lintel

84. Which order of levers is shown in the figure?

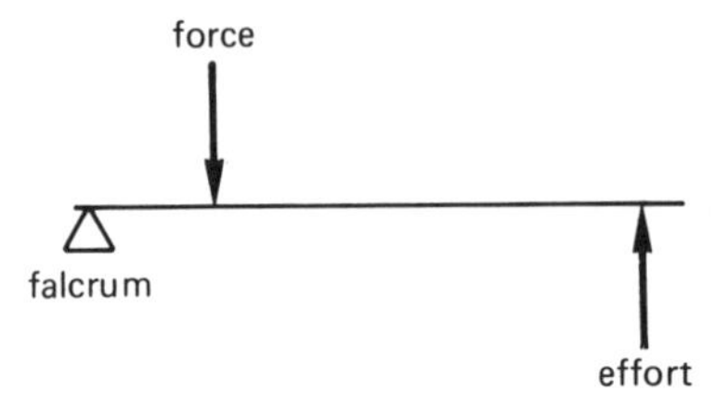

(a) first
(b) second
(c) third
(d) fourth

85. Ignoring the self-weight of the simple lever shown in the figure, what force is needed at x for equilibrium?

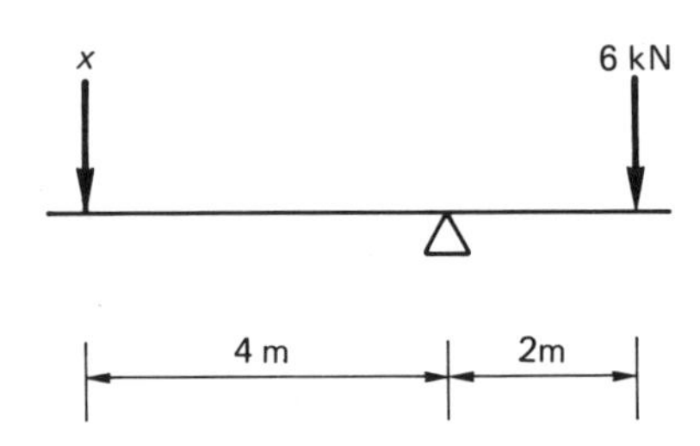

(a) 2 kN
(b) 3 kN
(c) 4 kN
(d) 6 kN

86. The use of a boat level is desirable when:
(a) using a long straightedge
(b) levelling individual bricks
(c) tamping oversite concrete
(d) checking small quoins for plumb

87. To remove a brick from an existing wall, it is necessary to use a lump hammer and a:
(a) plugging chisel
(b) raking out pick
(c) brick bolster
(d) tiling chisel

88. A long, timber straightedge used to compact concrete is called a:
(a) punner
(b) ram
(c) tamp
(d) vibrator

89. When clay subsoil becomes wet and freezes, foundations may be disrupted. This is known as:
(a) frost heave
(b) settlement
(c) subsidence
(d) foundation failure

90. The setsquare used for isometric drawing contains three different angles. These are:
(a) 45°, 45°, 90°
(b) 30°, 30°, 90°
(c) 30°, 60°, 90°
(d) 30°, 45°, 90°

91. Corner profiles for setting out a building may be constructed from:
(a) three pegs and two boards
(b) two pegs and two boards
(c) three pegs and three boards
(d) two pegs and one board

92. The type of foundation normally used on a sloping site is a:
(a) stepped foundation
(b) deep strip foundation
(c) short bored pile foundation
(d) continuous foundation

93. When a trench is excavated the spoil increases in volume. This is known as:
(a) expansion
(b) bulking
(c) swelling
(d) surplus spoil

94. Wire cut bricks may be recognised because they have:
(a) scratches on the face
(b) shiny surfaces
(c) no frogs
(d) deep frogs

95. A suitable foundation for domestic buildings built on shrinkable clay subsoils is a:
(a) raft
(b) wide strip
(c) stepped
(d) short bored piled

96. Extra wall ties are necessary in cavity walls of domestic buildings when the:
(a) house is to be built on reclaimed land
(b) building is to be over three storeys in height
(c) cavity is in excess of 75 mm wide
(d) building is to be heavily loaded

97. To prevent displacement of struts in trench timbering, the following are used:
(a) lipping pieces
(b) waling boards
(c) puncheons
(d) page wedges

98. Curing concrete refers to:
(a) Correcting faults on the surface
(b) adding extra water to the mixer
(c) keeping it damp while hardening
(d) regularly checking the water content

99. The number of bricks required to carry out the efflorescence test according to BS 3921 is:
(a) 10
(b) 20
(c) 24
(d) 36

100. The term 'batching' of concrete materials refers to:
(a) mixing
(b) measuring
(c) pouring
(d) transporting

ANSWERS TO MULTIPLE CHOICE QUESTIONS

1. (c)
2. (b)
3. (d)
4. (b)
5. (c)
6. (d)
7. (b)
8. (a)
9. (a)
10. (b)
11. (d)
12. (c)
13. (c)
14. (c)
15. (b)
16. (b)
17. (d)
18. (b)
19. (d)
20. (a)
21. (b)
22. (d)
23. (b)
24. (c)
25. (a)
26. (b)
27. (d)
28. (b)
29. (c)
30. (d)
31. (d)
32. (c)
33. (b)
34. (a)
35. (d)
36. (b)
37. (a)
38. (c)
39. (b)
40. (d)
41. (b)
42. (b)
43. (a)
44. (a)
45. (b)
46. (c)
47. (c)
48. (a)
49. (b)
50. (b)
51. (d)
52. (c)
53. (a)
54. (b)
55. (c)
56. (d)
57. (d)
58. (d)
59. (a)
60. (a)
61. (b)
62. (a)
63. (c)
64. (b)
65. (b)
66. (d)
67. (d)
68. (d)
69. (b)
70. (d)
71. (b)
72. (d)
73. (a)
74. (b)
75. (a)
76. (d)
77. (d)
78. (d)
79. (b)
80. (a)
81. (c)
82. (c)
83. (a)
84. (b)
85. (b)
86. (b)
87. (a)
88. (c)
89. (a)
90. (c)
91. (a)
92. (a)
93. (b)
94. (c)
95. (d)
96. (c)
97. (a)
98. (c)
99. (a)
100. (b)

INDEX